Fracking 101

Eric George BSc. M.Phil

Eric George

Fracking 101
Copyright © 2016 by J.E. George
ISBN: 9781532829727

Cover design by Jacqueline George
All cover art copyright © 2016 by J.E. George
Cover images courtesy Nilfanion & Joshua Doubek
Printed and bound in Australia.

PUBLISHER
Q~Press Publishing

Contents

Foreword

This book is a joint effort. Eric George worked in the oil industry and knows what he is talking about, but he doesn't write well. In fact we are thinking of marketing his writing style as a cure for insomnia.

Jacqueline (me) is the writer and although I deny being a complete bubble-head, I admit to being pretty well useless at oil wells. My job was to take the engineering text and ideas, and turn them into something that normal folk can read and understand easily. Between the two of us, I hope we have produced a book that is both factual and readable.

Writing about fracking is not simple. A little like an aeroplane, a fracked well is a complex construction made up of components that all have to work perfectly all the time, and you need to spend years on the rigs and in university to truly understand them all. But that doesn't stop anyone who really wants to see the bigger picture. With a little thoughtful reading, this book should allow you to stand your ground in any discussion over fracking.

I have to be philosophical about using metric or American units of measurement. Americans are usually annoyed by kilograms and metres, but the rest of world rolls its eyes when faced with something as difficult to understand as a foot. Really! We nearly all have two of them but still ask – how much is that in centimetres?

This is book about engineering and you can't avoid measurements, so I have tried to give equivalents.

Thanks are due to Chris Kalyta for his insights and technical editing, and to Debbie Shaw for her very pointed copy editing..

Lastly, don't worry too much if you dislike engineering or science. There are concepts in the book that might be difficult to understand so my advice is – don't try. It will be enough if you concentrate on the big picture and take the concepts for granted. So – read on, and enjoy the trip!

Jacqueline George
Cooktown 2016

Introduction

Fracking has become a hot button issue, lining up beside topics like climate change, gun control and, in the southern States at least, evolution. That's sad, because it is very much simpler than any of those but still shares the uninformed shouting that passes for public debate.

This book is an attempt to inform, to pass on details of the technique so anyone – for or against – can argue their case based on an understanding of the facts. I won't say I don't care what position you take but recently I have been really, really annoyed by the media, by friends and by the proverbial man in the pub who all give the world the benefit of their opinions at maximum volume, and are just making themselves look stupid.

Say what you want, but at least take the time to understand what you are talking about.

As with the other topics above, you should not take anything at face value. You have to ask just who is writing, and are they representing any vested interest. So to come clean, I - that is Eric George – used to work for an international oilfield service company, one of the two big players in the business. At the very end of last century I was made redundant for the crime of being over fifty in a year of falling oil prices. I felt bitter at the time, but that was fifteen years ago and I have to admit the game has always been played that way. Why keep a fifty year old around and fire the younger engineers you will need in ten years' time when the oil price recovers? Now I live in Queensland, on the shores of the Coral Sea, and you would not get me near an oil well with wild horses.

My line of work was mostly high pressure pumping into oil wells. Pumping cement to set the well casing in place, pumping acid and chemicals to clean up wells and, very occasionally, pumping frack jobs.

Counting on my fingers, I was involved in frack jobs in six

different countries. Always as part of a team, sometimes humping pipe, sometimes operating pumps, sometimes designing and supervising. So I do know what I am talking about.

Am I pro or anti fracking? I have to say, that's not a good question. Fracking is a technique that can be used well or badly. It's a bit like asking if you are for or against axes because axe murders are particularly gruesome. Banning axes is not practical, and the axe is not the root of the problem.

Fracturing is just a technique. Get it right and you can turn an impossible reservoir into a practical source of oil and gas. Get it wrong, and you can mess up the environment. I have to add here, get it wrong under a proper regulatory framework, and you will be forced to fix things before the damage is done – but we will talk about that later.

The fact is that a bad well – fractured or not – can cause a lot of damage. On the other hand, a well-designed, constructed and operated well can be a great benefit to mankind.

I want to help the debate along and the best way to do that is to give everyone a good understanding of fracking and possible problems. I hope this book will help.

Hydrocarbons and where they live

The first step in forming an opinion on fracking is to understand what we are talking about. That sounds like an obvious statement, and so it is. Fracking is a public issue and – in most countries – anyone can say what they like about it. However, if you want to influence public policy, you had better have some grasp of the techniques involved, and their implications. You can't argue a case unless you know what you are talking about, and you can't counter your opponents' arguments unless you can pick if they are telling the whole truth or leaving some dirty laundry in the basket.

I'm sorry, but this is going to involve a little serious reading. I will try and make it as painless as possible. I know some people have an allergy to maths and technological detail, so let's try and do as much as we can with just descriptions and pictures. If you don't understand something the first time around, that's normal. Go back and read it again, and you will soon be on track again.

You may find it helps to have a 1 ft (yes – 30 cm) ruler on the table as you read. You will see why in a minute.

Fracking – more properly hydraulic fracturing – is one of the last steps in the making of an oil or gas well and it does not make any sense to look at it in detail until we have covered the basics of well siting and construction. As we go along, you will find that most well problems are created during construction and we have to understand the basics of well drilling before we can identify the problems. Let's begin with some fundamentals.

Hydrocarbon Basics

Oil and gas wells are drilled to produce oil and gas, but rarely produce just one or the other. They nearly all produce a mixture of the two, and they nearly all produce significant amounts of water as well. To save messing around, we call them all fluids – oil, gas, water, mixtures – they are all called fluids. You may hear produced fluids referred to as hydrocarbons; wells produce hydrocarbons which we want, and water which we generally don't want.

There is a good reason for referring to produced oil and gas simply as hydrocarbons – because the distinction between the two is not always simple. The gas you use at home is methane and this is known as 'dry gas'. Methane is essentially odourless so energy companies add a smell to it, to help us know if there is a leak in our house. Gas coming naturally from the well may include 'wet' fractions – ethane, propane, butane and others. They are also gaseous at normal temperatures and pressures and are much more valuable than straight methane. (Don't try to remember all these names – just remember the idea of 'dry' gas with various 'wet' components.)

The next fluid produced, and usually the most valuable, arrives on surface as a volatile liquid known as condensate. This is as rich as or richer than gasoline, needs little refining, and is used in the refinery to blend into other liquid products.

And then we have all the various liquid mixtures that make up crude oil. Plus water. There is always water to make everyone's life difficult, because water that has been trapped underground for geological time frames (i.e. tens of millions of years) is inevitably salty and hard to dispose of economically.

Just bear in mind that when you see the word 'fluid' below, it refers to both gases and liquids.

It is possible and useful to frack some oil wells, but nearly all modern fracking is for gas production. Just hold on to the idea that fracked wells normally produce hydrocarbon gas, along with some hydrocarbon liquids (welcome) and some water which always gives trouble to everyone.

Geological Basics

Hydrocarbons are formed when natural sediments with a percentage of organic material are buried by geological processes and subjected to high pressures and temperatures.

That's it. Find yourself some river mud or undisturbed sea bed, bury it several kilometres deep, wait a few million years, and the earth's natural heat combined with the weight of overlying sediments, will cook it up to form rock and hydrocarbons. The richer the original mud was in organic material, the more hydrocarbons will be formed.

The rock stays where it is, of course, but the hydrocarbons are naturally mobile and free to move around. They commonly migrate out of their source rock and can go anywhere. The gases are the most mobile – confining high pressure gas in one place needs just the right geological conditions. Most hydrocarbon gas leaks up to the surface and into the atmosphere. It leaks at a very low rate, but it does so over extremely long time periods and this will deplete most source rocks without us ever knowing about it.

The oil and gas industry is built on the pockets of volatile hydrocarbons that did not escape. They flowed from their source rock laterally and vertically (it is nearly always easier for hydrocarbons to move sideways rather than vertically) until they layers of very tight rock trap them. There they sit, escaping only very, very slowly (because no rock is completely gas-proof.) The rock holding the hydrocarbons is known as the reservoir. The confining rock above the reservoir is known as the cap rock or seal.

In some places around the world – the Alberta tar sands, for instance – all the beautifully profitable volatiles have escaped and left only a tarry residue which is difficult to exploit.

To complicate the picture a little, there are regions of the world with very thick bodies of impermeable fine-grained rock containing an organic component. Of course, they have generated hydrocarbons – and may still be doing it. The fine grained rocks we are talking about here are shales, and shale formations can be *huge*. Huge reservoirs with, of course, huge quantities of gas – trapped in rock it cannot easily flow through.

So add source rock, reservoir and cap rock to your notes. You will meet them again.

Permeability

A permeable rock allows fluids to pass through it easily. An impermeable rock stops them passing through.

Suppose you pour a bucket of water onto a sand dune; it will disappear very quickly. Do the same thing on a clay tennis court and you will be standing in a puddle. The water will eventually dry up or run away. Only the smallest fraction will soak in. This is the difference between permeable and impermeable rock.

The Middle East has been blessed with oil reservoirs that are not only thick and very extensive, they are also extremely permeable. Lucky them, because no-one else can produce oil so easily and cheaply. They will be producing oil long after the rest of the world's oilfields have run out.

The Middle East, and Russia in particular, also have large gas fields with relatively permeable reservoir rock, far more than their own economies can use. They have surplus gas, and they can easily send it to Europe through pipelines, or compress it to liquid and send it by ship to countries like Japan.

That is the easy gas. Much more common are the difficult shale reservoirs. Shale is formed from mud or clay with little in the way of sand. It is usually laid down in deeper water as a muddy sea bed with a small fraction of organic material. Over geological time (tens of millions of years), the mud is buried, compressed and cooked to form shale, and the organic fraction goes to form gas.

A hundred years ago, the presence of shale gas was known about, but no-one cared. Gas flows so slowly through shale that wells drilled into shale were abandoned as useless.

Starting to Drill

When and where the first oil wells were drilled is a subject for Trivia Nights at your local club (or pub). Certainly oil seeps and tar pools have been known and used for millennia. Shallow wells were being drilled in China in the 4th century AD, but the first combination of wells and a refinery was probably built by Fiodor Priadunov in the Ukhta region of the Komi Republic in Russia in the mid-18th century.

A century later, oil was an inconvenient by-product of the wax mining industry in Eastern Galicia (parts of modern Poland and Ukraine) until it was refined for lamp-oil and wells were actively drilled to increase supply.

Shortly afterwards came the first commercial well in North America (I don't dare get into the argument between the US and Canada about who was first), and the oil industry had found its natural home. Its expansion was meteoric.

Why did the oil industry become so big, so quickly in North America? Strangely enough, the answer lies not so much in the geology as in a quirk of the legal system. In Europe, all land was and still is owned by the monarch or state. A person might think they own a farm or a house and garden, but in fact they just have the 'freehold' – they hold the land from the monarch or state without obligation.

A legal nicety, except just about every country limits a landowner's control of his land to the surface. You might be permitted to dig a hole a metre or two deep, but go any further and the King or President will be very upset.

In Europe, all the oil and gas belongs to the government, and they do deals to allow big companies to exploit it with big projects.

In the States, things are different. There the landowner may own his land *all the way down* and any oil or gas belongs to him.

An individual will see a low producing well as the goose that laid the golden egg. A big company will see it as no more than a nuisance.

So small wells are not common in Europe, but there are many, many thousands of them in North America. It follows that the technology for extracting and producing oil and gas was developed by the people doing most of the work i.e. North Americans. This is an important feature of the oil industry because it means that around the world the majority of wells are drilled and put into production using North American techniques and equipment, and engineering standards are based on those developed in America.

It also means that writing a book like this is a pain, because American engineering is alone in not using the metric system. The industry has to adapt and engineers in Germany might drill a hole to 4000 metres (~12,000 feet) deep, but its diameter will be 8-1/2 inches and the casing they run to bottom will be 7 inches diameter, because those are the sizes of the drill bits, casing and all associated handling equipment.

Oh well, they live with the measurement problems and I guess we will have to do the same. I will try and remember to give rough equivalents where it matters. For when we are speaking approximately, please assume three feet amount to about one metre, and a US gallon is pretty much four litres.

The life of a well

The well is the visible sign of a hydrocarbon project, and, at this point, it is worth looking at its environmental impact on the well surroundings.

The location in the landscape might simply be a matter of positioning it directly above the most attractive spot in the reservoir, and this is common where land values are low. Where land is valuable – say in an urban environment – it is much more common to set up a single location for several wells and drill angled holes from the one spot. This way a large section of a reservoir can be covered from one base.

A well goes through four distinct processes during its lifetime. The first is the actual drilling. This is the noisiest and most exciting period, with an intrusive derrick covered in lights, drilling away twenty-four hours a day. There will be all sorts of associated equipment, either fixed or driven on and off site as required, and lots of people. There is no missing a well if you are unlucky enough to have one drilled next to your home.

The good news is that this phase does not last very long, typically four to six weeks, although it could be more or less depending on the depth and drilling difficulty.

Once drilling has finished, the site is virtually cleared, usually leaving no more than the well-head valves (the Christmas tree) and some buried pipe-work. Most of the ground used during drilling is released back to its former use, but access is needed for daily monitoring and for occasional service work. Service work may need a light work-over rig, which has a far smaller footprint than a drilling rig.

Fracking is usually scheduled immediately after completion of the well and before it is put into production, but it can also be done later. Service work can involve replacing or servicing surface and downhole equipment, or cleaning and stimulation. Sometimes, a small fracking operation is used to clear damage around the well-bore and provide easier access for the produced fluids. These sorts of operations are infrequent and are normally over in a day or two.

For most of a gas well's life, it will sit and quietly produce. A passer-by does not know if it is working, or shut in. A low volume, shallow oil-well might be fitted with a nodding donkey pump; gas rich wells will have nothing.

The final stage of a well's active life is reached when

production falls so low that its value does not match the costs of production. When will this happen? It depends firstly on the reservoir, but to some extent on the well itself. A properly constructed well will last longer than one drilled on the cheap.

Once the accountants write a well off, a service rig will move onto site and remove the down-hole equipment – tubing, packers, valves etc. Then it will pump liquid cement into the well, ideally filling it back to surface. Some jurisdictions are used to skimping on this process, and that can store up trouble for the future.

A thing that people not in the business often forget is that hydrocarbon extraction is a mining activity. It is not like a factory that can produce cars into the distant future, changing models, staff, and suppliers as their business changes. You drill a well, produce from it until the hydrocarbons run out, and then you dismantle it. A responsibly abandoned well leaves nothing above ground. After a year or two, you'd never know it had been there.

The damage left by a properly managed well should be invisible and much more acceptable than that left by a coal mine or nuclear power station.

Drilling a well

During the last century there has been very little change in the basics of oil and gas well construction. A driller from 1930 could be put on the floor of a normal land rig and, after a small amount of familiarisation, go straight to work. There have been big changes in control and monitoring systems but the basics of drilling a hole and setting casing have remained the same. Let's suppose we are going to drill to 10,000 ft (~3000 m) and look at the basic steps:

The hole is drilled using a bit mounted on the end of a pipe – it's as simple as that. Once the hole has been drilled, steel casing is set in it, and the well can be completed ready to go into production.

Of course, things get a little more complicated than that, so let's take it step by step.

The first section of a typical well, nearest to the surface, might be drilled with a diameter of 30" (~750 mm) and go down 100 or 200 ft (~30 – 60 m). It is drilled in the softer, less stable surface soil and rock. It would be impossible to simply drill a hole, as it would inevitably begin to collapse as you drill. To counter this, drillers use mud.

Mud is a crucial component of well construction and does several important jobs. It is pumped down the drill pipe and out of the bit, cooling and lubricating as it goes. The mud is mixed to a controlled viscosity, high enough to stop it leaking off into the rock and to allow it to transport the drill cuttings back to the surface. (Drill cuttings are the rock fragments, sand and fine materials made by the drill bit cutting into the rock.) Mud also coats the walls of the hole, which minimises leakage into the formation and allows the hydrostatic pressure to act against the walls of the hole. This pressure is essential to keep the hole open and stop it falling in on itself.

When a diver goes deep under the sea, he is subject to the weight of sea water above him. This pressure is known as hydrostatic pressure and is a key part of drilling a hole. The hydrostatic pressure of the mud in the hole supports it and stops the walls collapsing in around the drill pipe. You can imagine the hydrostatic pressure at the bottom of 200 ft of heavy mud is quite high. At the bottom of a 2000 ft hole or deeper, it is very high indeed and we will have to think about the implications of that later.

Back to the hole we have just drilled – say 200 ft of hole full of mud. The next step is to run steel casing into the hole and cement it into place. For 30" holes, the normal casing diameter is 20" (~500 mm). The casing comes onto site as 40 ft (12 m) sections which are screwed together to form a continuous lining for the hole. Getting it into the hole is easy; cementing it in place is a different issue so let's deal with that now.

Cementing casing into place

A proper bond between the casing and the rock formations is vital to the well. It not only supports the casing but it also isolates the different rock formations the hole has penetrated. Suppose the well passes through a water bearing layer (or aquifer) that is used for human consumption or agriculture. The last thing we want is for that precious water to leak into permeable layers down below, or be contaminated by salt water or even hydrocarbons from higher pressure layers at depth. The only way this aquifer can be protected is by providing a good quality, continuous sheath of cement around the outside of the casing.

Cementing operations become very difficult and complex further down the hole, but cementing the surface casing will demonstrate the basics of the process.

We start off with an open hole, full of drilling mud:

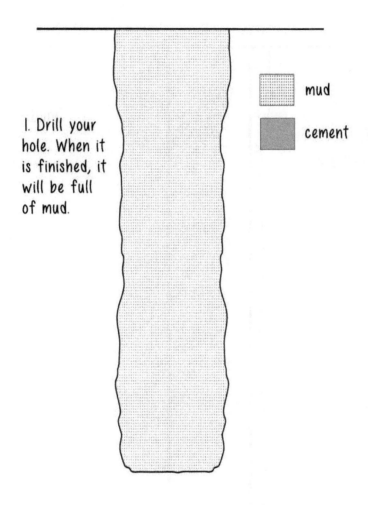

1. Drill your hole. When it is finished, it will be full of mud.

mud

cement

Our next step is to run casing to the bottom of the hole. At this stage the casing is full of mud, as is the space around it. This space is known as an <u>annulus</u>, a term you will come to know well as the hole gets deeper.

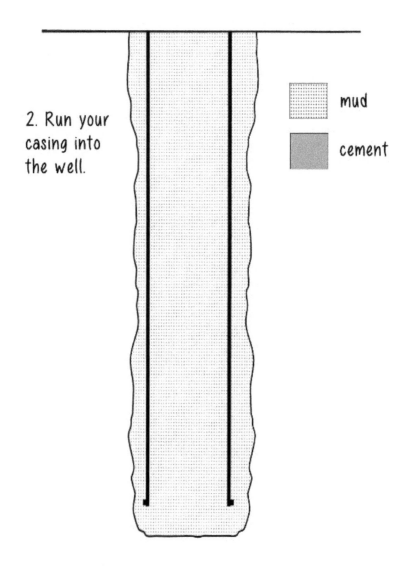

2. Run your casing into the well.

mud

cement

Now we start cementing. A liquid cement slurry (basically cement powder and water) is pumped into the casing, followed by mud.

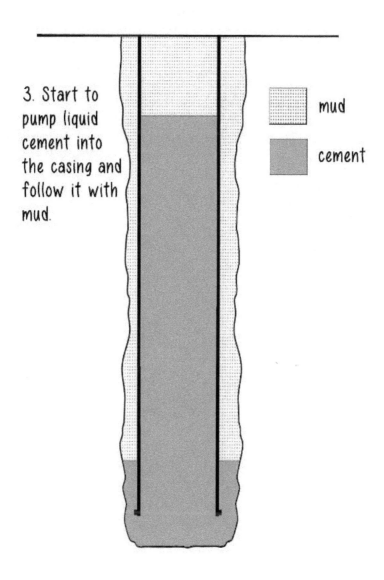

3. Start to pump liquid cement into the casing and follow it with mud.

mud

cement

Next comes the critical stage. More mud is pumped into the casing, forcing the cement slurry out of the bottom and back up the annulus. Once the slurry reaches the surface, pumping stops, all valves are closed and the cement is allowed to set. The casing is now sealed into the well and drilling deeper can start.

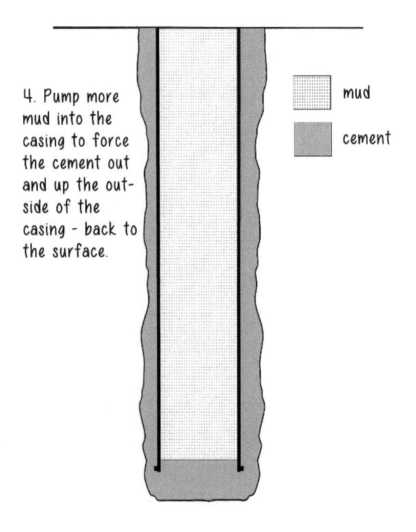

4. Pump more mud into the casing to force the cement out and up the outside of the casing - back to the surface.

mud

cement

Drilling ahead...

So – where have we got to? The well has been drilled through soft unstable surface material, and the <u>Surface Casing</u> has been cemented into place.

The next step is to drill ahead as far as we safely can and set the <u>intermediate casing.</u> We need to run a drill bit to the bottom of the hole, through the surface casing which is normally 20" (~500 mm) diameter. The standard diameter for the deeper hole is 17-1/2" (~445 mm).

How deep will the intermediate hole go? That depends entirely on the formations the hole is being drilled through. Perhaps experience in the area around the site tells the well engineers that an unstable formation can be expected at around 1500 ft. In that case, every effort will be made to drill past it and set the next casing across it. If the unstable formation is safely sealed off, it will not come back to haunt you as the hole gets deeper. For a well like ours – 10,000 ft deep – the intermediate hole would commonly go to 1000 – 2000 ft.

Generally speaking, the deeper you drill, the harder and more stable the formations become but the intermediate section of the hole can expect to encounter various difficulties (we will discuss those as we go).

As the hole is drilled a large volume of mud is prepared on the surface and pumped down the drill pipe. It returns to the surface in a constant stream, bringing with it all the drill cuttings. On the surface it is run over large vibrating mesh tables, which separate the cuttings from the mud which is returned to the mud tanks and re-used.

The Drilling Mud

As we drill deeper, the physical properties of the drilling mud become more and more important. We will look primarily at the <u>weight</u> of the mud and its <u>viscosity</u>.

As before, the mud is lubricating and cooling the bit, besides transporting the drill cuttings back to surface. The mud also coats the walls of the hole and prevents liquid leaking off into any permeable formations. And the hydrostatic pressure exerted by the column of mud pushes back against any formation fluids and keeps them from flowing into the well.

The lists of tasks above is a lot to expect from something that is essentially no more than muddy water so you are not going to be surprised to hear that mud is actually very complicated. Let's start by looking at its weight. If it is too light, its hydrostatic pressure acting on the hole walls will not be enough support weak spots and the hole will start to collapse.

Being too light can also allow formation fluids into the well. Hitting a permeable layer will be an adventure because it will certainly contain water and that water is going to rush into the well and may flow all the way to surface. You have now lost control of the well and you may be in big trouble.

Of course, trouble with an inflow of water is insignificant compared to a break-in of gas. Shallow pockets of gas are common in some areas and touching one is like popping a champagne cork. Near surface gas can reach surface very quickly and catch a driller by surprise – very dangerous. And the gas may not be a hydrocarbon. It could also be hydrogen sulphide (H_2S) and that is a deadly poison.

To control fluid entry, mud weight is increased. This is done by mixing in some weighting material – usually powdered barite. (Barite is barium sulphate – a naturally occurring mineral, used because it is inert, heavy and cheap.)

Increasing the mud weight to control formation fluids does bring another risk. If your mud is too heavy, the pressure on the hole walls may become too high and your mud will start to leak off. The leakage can grow to problem levels, even reaching the situation when most of the mud pumped into the drill pipe just disappears and does not return to surface. This situation is known as 'lost circulation' and will stop you drilling ahead. You will have to pull the drill string back part way and try to re-establish circulation, perhaps by adjusting the mud weight and/or adding some sort of plugging material to it.

The next important property of the mud is its viscosity. If the mud is runny, it can be circulated around the well easily. If it is too runny, it may leak off easily or have trouble transporting drilling cuttings back to surface.

If the mud is too thick, excessive pump pressure will be needed to circulate the well, and this pressure is also applied to the bottom part of the well – encouraging leak off or even lost circulation. This can be a problem in the intermediate hole, but the problem gets bigger as the well gets deeper and the hole diameter smaller.

As the drilling mud picks up more and more fine material at the drill bit, it naturally becomes heavier and more viscous. To counter this, at the surface the mud is not only passed through finer mesh sieves, but through vortex cleaners as well. In some conditions, especially in deep holes where there is less margin for viscosity and weight changes, large centrifuges may be added to the cleaning equipment.

Additives

The preparation and maintenance of a mud system needs at least one full time specialist on the drilling rig, together with help from the drilling crew. As the well gets deeper, more and more additives are used to change and control mud properties, and these additives are worth a book in themselves. For the moment, just bear in mind that 'drilling mud' is a sophisticated mixture of liquid, clay and various chemical additives.

So how are we doing?

By now, our intermediate hole should be reaching its full depth, and we have, say, 1500 ft (~500 m) of 17-1/2" (~445 mm) hole full of drilling mud. We have drilled through a variety of rocks, none of them interesting enough to hold hydrocarbons, and have encountered day-to-day difficulties that we have overcome. What next?

We are drilling above a known hydrocarbon reservoir or perhaps above what we hope will become a known reservoir. We should take the opportunity to collect as much data as we can that might help in the future. It is time to call the geophysical logging truck onto site. (This sort of logging truck has nothing to do with chopping down trees!)

Geophysical logging is an extremely sophisticated business based on lowering electronic tools into the well on a strong but flexible steel cable with signal cables at its core. It aims to collect as much data as possible on the rocks around the well. The tools are lowered into the ground and as they are slowly recovered, a long chart is generated showing how the formation's physical properties change with depth.

Many different tools and techniques are used for the sections of the well likely to contain hydrocarbons. That does not include our intermediate hole, so a simple array of tools will be used. We will probably be looking only for changes in the way it transmits electricity (resistivity, SP) and the physical size of the hole. Resistivity and SP provide a simple signature that we might need to compare with future wells in the same area.

The physical size of the hole is important, because we have to run casing and cement it into place. Intermediate casing is traditionally 13-3/8" in diameter (~ 340 mm). 13-3/8" casing in a 17-1/2" hole leaves a relatively large annulus between the casing and the hole wall which must be filled with cement. Depending on the setting depth, the 13-3/8" cementation is often the largest cement job on a well, requiring many tons of cement.

The cement job follows the same procedure as used to cement the surface casing and here is a picture of the well once the job is done (a section of the well has been omitted so the picture will fit onto the page):

So there we are. Your well has been drilled out, the intermediate casing has been run, and you have cemented it into place. Except - it's not that simple. Well, I did warn you that the deeper you go, the more complicated everything becomes...

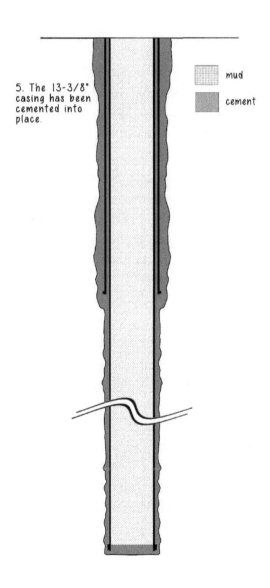

Deeper, Heavier, and the Frack Gradient

Pressure gradients

Let's go back, and think about a well full of water. The column of water exerts a pressure against the wall of the hole, and that pressure gets higher the deeper you go. The principle is no different to diving into the sea. If you snorkel down to 10 ft (~3 m), you can feel the weight of water pressing against your body.

If you put on some scuba gear, you might want to go three times deeper and you will definitely be aware of the pressure on your ears. Go deeper still and funny things start to happen to your body. The nitrogen in the air you are breathing begins to collect in your blood, and you will need to follow a de-compression schedule as you slowly return to the surface.

That is the principle of hydrostatic pressure, acting on a very small scale. With a well, the pressures are very much higher because the depths are much greater. We can calculate the pressure at any point in a well by considering the <u>pressure gradient</u> of the fluid filling it.

This is an easy concept to grasp. Plain water has a specific gravity of 1. That is 1 litre of water weighs 1 kilogram. In non-metric units, one US gallon of water weighs 8.32 pounds. These are the base figures we start with.

Take it from me (and don't bother trying to calculate it), a foot depth of plain water exerts a pressure of 0.433 psi. You pump your car tyres up to pressures of around 30 psi, or the pressure you get at a depth of about 70 ft.

In metric units, a metre of water gives a pressure of around 9.8 kilopascals and your tyre pressures are around 200 kpa.

So, the pressure exerted by a column of water is 0.433 psi/ft of depth. Just take the depth in feet, multiply by the <u>pressure gradient</u> of water (0.433 psi/ft), and you will have the hydrostatic pressure at that depth.

When you get to the bottom of a well, this figure has become very large. A 10,000 ft well full of water will have a bottom hole pressure of

4330 psi – which makes your car tyre pressure look tiny.

That pressure was for water – a light liquid weighing 8.32 ppg (pounds per US gallon). In practice, by the time you reach 10,000 ft depth, you will be drilling with a significantly heavier mud. If we suppose your mud weighs 11.5 ppg (a specific gravity of 1.38), then the hydrostatic pressure at the bottom of your hole will be 5970 psi.

The hydrostatic pressure at the bottom of a well is high, but does that matter? Well, yes and no. The well has been drilled into rock and that is capable of resisting great pressures. On the other hand, what would happen if, in addition to the hydrostatic pressure, we started to apply extra pressure by pumping into the well?

Suppose we close in the well at surface so nothing can escape and begin pumping mud into it. The pressure in the whole system will start to increase. Pressure gauges at the surface will show 500 psi, 1000 psi, 2000 psi…

At the bottom of the well, the rise in pressure is the same but added to the hydrostatic pressure already there. Our well above with a hydrostatic pressure of 5970 psi will now have a total bottomhole pressure of 7970 psi.

The first response from the well will be increased leaking. The mud will be trying to escape and any weak or permeable formations will be under attack.

There is a limit to how much pressure can bleed off this way. The static pressure in the leaky formations will be increased as mud is pumped into them until they can no longer accept more mud. The pressure in the well will continue to increase as the surface pressure rises.

Of course, this cannot continue indefinitely and, in the end, something will break. And it will be the rock itself. It will literally break open and allow mud to flow into it.

This is hydraulic fracturing. High pressure is applied to a rock formation until it breaks. The process is as simple as that. We will go into the details of exactly how the fracture happens later but, for the moment, let's just note that if the pressure in a well is increased beyond a certain point, the rock will break.

The Frack Gradient

What pressures are we talking about? Surprisingly, the pressure needed to fracture a rock formation is quite predictable. No matter what sort of rock we are talking about or its depth in the well, fracturing will happen at a pressure gradient of 0.7-0.75 psi/ft. It's as simple as that. For a 10,000 ft well, if you apply a pressure at the bottom of about 7500 psi, you will start a fracture. And that is the same if we are talking about shale, sandstone or even granite.

Drilling ahead – again

We have set our intermediate casing (13-3/8" or 340 mm) and it is now time to drill ahead. The next hole size – determined by the biggest drill bit that will fit through the intermediate casing – is 12-1/4" or 311 mm. We will drill this hole to a position just above the expected reservoir and run 9-5/8" casing. This is known as the production casing.

Working out the depth to set the casing, and understanding the dynamics of drilling the hole and running the casing string, is a job for the drilling engineers. It takes a great deal of computer time and sophisticated modelling to work out how or even whether the casing can be set where the geologists want it. Fortunately, I just did this in my head and came up with a setting depth of 9500 ft...

More trouble ahead – cementing at depth

We do not need to look at the increasing difficulties of drilling as the hole gets deeper because they do not affect later fracturing operations. We will let the drillers and engineers get on with their jobs while we take a look at one of the critical steps in well construction – cementing casing in deeper holes.

Wells use Portland cement – the same material we use on the

surface to make reinforced concrete, roadways, and bridges. It has been around since the 1850s and we know a lot about it. It is cheap, reliable and strong, and ideal material to bond well casing into a drilled hole.

We have seen how a slurry of cement and water can be pumped into a casing string and back up the annulus between the casing and the hole wall. Once it has set, the casing is effectively part of the surrounding geology.

This sounds easy enough, and so it is for the surface casing. It is not much more difficult for the intermediate casing, although here you may need some added chemicals to slow the setting process and give you enough time to pump it into place, along with additives to make it runnier and easier to pump. You will probably have to start worrying about the weight of the cement slurry and the hydrostatic pressure it will give.

By the time your hole has been drilled to 9500 ft and you are cementing 9-5/8" casing into a 12-1/4" hole, things have become very, very much more difficult. Let's run though the things the cementing engineer worries about and how the problems can be handled.

1. **Pumping and setting times**

   As soon as water is added to cement powder, it begins to set. There is a period when cement slurry is liquid enough to be pumped, a period when the setting mixture is unpumpable but has no strength, and then the time when the chemistry begins to work and crystallization means the slurry has become a true solid. This solid will rapidly gain strength and then the rate of strength improvement will drop off. On surface, most of the strength gain will happen in the first week, but strength will slowly increase into the distant future. Downhole, things happen much faster, as we shall see.

   To mix and pump the cement slurry for the 9-5/8" casing might take 4 hours, so the cement engineer must add chemicals to the slurry to give him that much pumping time plus a good safety margin in case things go wrong.

The cement slurry is mixed with water plus a range of very important additives. Perhaps the most basic one is a retarder which slows down the setting process and allows the slurry to be pumped for longer – how much longer depends on the temperature the slurry experiences.

What is the temperature at the bottom of an oil well? It increases the deeper you drill and we can say the static temperature increases by 25 to 30 ° Celsius/1000m [~15 °Fahrenheit /1000 ft]. We can expect the static temperature at the bottom of our 3000 m (10,000 ft) well to approach 100 °C (201 °F). In practice, just circulating the well with mud will give a dynamic reduction in bottomhole temperature and the cementing engineer will take this into account when the slurry is designed.

2. Cleaning the Hole

On completion of drilling, the hole walls will have built up a significant cake of mud. There may also be large quantities of drill cuttings circulating or tucked away in washouts. If the cement is to form a good bond with the formation, the hole must be as clean as possible. This is done by running scraping tools into the hole before the casing is run, and by circulating the hole clean. Ahead of the cement slurry, the engineer will normally pump cleaning chemicals to flush out the last of the mud and debris.

3. Centering the Casing

Although the casing is hanging in the hole before cementing, you must not think it is automatically hanging centrally. Far from it. The clearance between 9-5/8" casing and the 12-1/4" hole is only some 20 mm (~3/4") either side, and the slightest displacement of the hole will mean the casing is actually lying against the hole wall. We want to surround the casing with cement slurry, something that is not possible if the casing is touching the hole wall. This problem is dealt with by using centralisers, springy steel bows set in a double collar that keep the casing away from the wall and allow mud, cleaning liquids and cement slurry to be pumped completely around it.

These centralisers are an essential part of a good cement job but they are often the cause of frustration on the rig itself. Drillers know it is much quicker and simpler to run slick casing without centralisers into the well. Centralisers hold things up on the drill floor and make life difficult. We will return to this problem later...

4. Static and Circulating Pressures

We have already dealt with the static pressure of a column of liquid in a well, and here there is an obvious problem. Our mud may have a bottom hole static pressure of nearly 6000 psi for mud weighing 11.5 ppg (SG 1.38 or 138% of the density of water). Full strength cement slurry (known as neat cement) is much heavier than mud, normally 15.8 ppg or SG 1.84.

If you are quick with your calculator, you will have worked out that the static pressure gradient of a column of neat cement slurry is already 0.82 psi/ft and if you look back to our section on frack gradients, yes, Houston, we have a problem. Frack gradients are in the range 0.7-0.75 psi/ft. The weight of a static column of neat cement is enough to start a fracture, even before we add the inevitable circulating pressure.

Circulating pressures are a significant problem. Imagine you are filling a bucket from a tap in your garden. Now fill the same bucket from the same tap but this time, through a long length of garden hose. It will take longer. In fact, if it is a long, long hose, filling your bucket will take ages.

The difference lies in the friction in the hose, the resistance to flow inherent in the liquid flowing through a pipe, and that varies with the viscosity of the liquid itself. (If you had been filling your bucket with honey, you may as well put it down and come back tomorrow…)

The effort of forcing cement slurry through the narrow annulus between the casing adds significantly to the bottom hole pressure and the cementing engineer is always struggling to minimise it and prevent accidental fracturing. Get this wrong and your cement slurry could disappear off into the surrounding rocks. That might not be too bad, except it will never disappear entirely. There will always be some left in the well bore to stop you re-establishing circulation. You will end up with the worst of all worlds, the casing being only partially cemented. Getting yourself out of that mess will be very difficult and you may even have to abandon the well and start over.

Cementing Operations

You might be wondering why, in a book about fracturing, we are spending so much time worrying about cementing the well casing into the hole. The answer is that the quality of the cement job is absolutely critical to the well. An inadequate job may mean the well cannot be fractured.

The cement bonds the casing to the formation and makes it part of the bedrock. It supports the pipe and prevents contact with possibly corrosive brines. Most importantly, it fills and isolates the annulus. This isolation is vital because it means well fluids – gas, oil, water – cannot sneak back up the well *outside* the well casing and contaminate aquifers or even break through to the surface.

Before we can really understand the difficulty of cementing a well, we need to have a good grasp of the scale of the problem. Do you have a street in your town that is 2.5 km (1.5 miles) long? On second thoughts, make it a bit longer as we have still to drill deeper once we have this casing cemented into place.

Suppose we lay a string of 9-5/8" casing along the length of that street. That is, a steel pipe with a diameter equivalent to your spread hand thumb to pinky plus two fingers widths. (If you are a big man, perhaps one finger width; a delicate lady might need three fingers). This is not a very fat pipe, but it is very long.

Let's set up your cementing equipment at one end but, before we start pumping cement slurry, let's walk (or drive!) down to the other end and consider the annulus we will be trying to fill.

For our current casing string, we have drilled a 12-1/4" (311 mm) hole. That is the size of drill bit used. Of course, we are drilling in natural rock so the hole is not gun barrel smooth and there will be places where the hole has broken to larger diameters, but let's work on the nominal 12-1/4" size, the common hole size for 9-5/8" casing. How big is the annulus around the casing that we want to fill with cement? Only 1.3" or 33 mm all around the pipe. That is equivalent to the width of your thumb when you press it on a table.

Let's think about that again. You are sitting at one end of the casing string preparing your cement slurry, and you are hoping to achieve a complete and regular coating of cement the thickness of your thumb, all around the pipe; 2.5 km away, where you can't see or feel or measure, in conditions of high temperature and very high pressure.

And it gets worse. Casing diameter is always measured as the outside diameter, but 9-5/8" (244 mm) casing comes in sections joined together with threaded collars, and the outside diameter of the collars is 10-5/8" (270 mm). That leaves you a space between the collars and the hole wall of only 0.8" or 20 mm.

Very few people understand the full drama of oil field cementing operations. They are technically difficult, using specialised pumping equipment and highly trained operators. They need strict quality control of the cement slurry as it is mixed, and they need to prepare and pump the slurry very quickly – commonly using 1.5 tonnes of cement powder per minute. To make life more difficult, they always seem to happen at 3 am, in the rain, but even in daylight the tension around the well site is tremendous.

It may have taken a month or so to drill our well to this point. We are now going to mix cement powder with water and all the various chemicals it needs, and pump it into the well. From that moment on, the process is irreversible. Sometime soon, the cement will inevitably set and if it is not where it is meant to be at the time, the well may be ruined. No-one is relaxed during a cement job; everyone is on edge, and everyone breathes a great sigh of relief when the pumps shut down with the cement sheath in place. Believe me, I still wake up remembering some nerve wracking times I had as a young cementing operator.

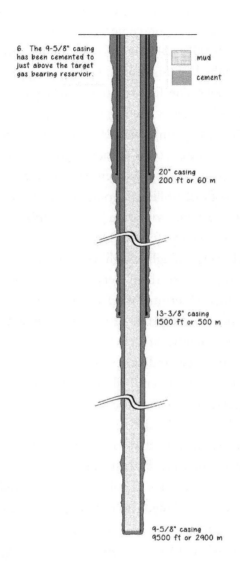

6. The 9-5/8" casing has been cemented to just above the target gas bearing reservoir.

mud

cement

20" casing
200 ft or 60 m

13-3/8" casing
1500 ft or 500 m

9-5/8" casing
9500 ft or 2900 m

The Top of Cement Problem

The cement sheath bonds the steel well casing to the surrounding rock, but, as we hinted above, there is an inherent problem. We are trying to cement the 9-5/8" casing into place at 8000 ft depth. The static pressure given by a column of liquid cement slurry plus the friction pressure of pumping it into place mean that we cannot pump

37

enough neat cement slurry to seal off all the 12-1/4" hole. The pressure of a full column of liquid cement would be high enough to fracture the rock and we would lose our cement, and possibly the well. Our particular well has 8000 ft (2440 m) of open annulus and perhaps half of it cannot be reached by this cement operation.

What can be done? The first solution is to lighten the column of cement slurry. This is done by keeping the best quality neat cement for the bottom of the casing string, and using lighter cement above. (The bottom end of the casing string is most critical because we will be drilling ahead and need a strong cement sheath to support the casing while the heavy work of drilling is going on. The cement above will not be subject to so much vibration.) We describe the two cement types as the light lead slurry, followed by a neat cement tail slurry.

There is a wide range of light weight cement materials we could use as our lead cement, some of them weighing as little as 5% more than water i.e. lighter than all normal mud weights. These cements are by far the best solution to the problem of cementing long well sections but they do have a significant drawback – they are expensive. Well construction is a business and expensive technologies are only used when either the environment allows no alternative, or when Government regulation demands them.

There are well known alternatives. Clay (bentonite) or fly-ash can be mixed with the cement. Both of these are cheap and easy to obtain but can only reduce the weight a certain amount – to a point perhaps 50% heavier than water. This weight reduction certainly helps the cementing engineer as he designs his cement jobs but do not allow him to solve our problem.

Another approach is to use a two-stage cementing operation. A special collar is run in the casing string and, after the first stage cementing has been completed, ports in this collar are opening by dropping a solid plug into the well and apply pump pressure to move a sliding sleeve. Once these ports are opened and the first stage has begun to set, a second cementing operation can be run to bring the top of the cement well up inside the previous casing.

This sounds complicated and, yes, there are lot of things that might go wrong with multi-stage cementing, and they frequently do. Two stage cementing is much less popular than it once was because we now have better alternatives.

Why is the top of cement position so important, especially for wells that are to be fractured? Because we do not want any open sections of well bore outside the casing to provide routes for fluids to move between formation layers either now or into the future when the well casing might corrode.

For our well, if we left the top of cement at, say, 4000 ft (1200 m) we would have a long section of open hole. Fluids in that annulus would be free to move upwards, either into the annulus between the 9-5/8" and 13-3/8" casing or – much worse – by-pass the 13-3/8" casing altogether and have access to shallow aquifers or even to the surface itself.

The Importance of Good Cementing

It cannot be stressed enough that a good cement sheath bonding the well casing to the surrounding rock is vital. This can be achieved with sufficient care. Modern monitoring equipment can show that it has been achieved or indicate problem areas.

In the bad old days (which are as far away as last week in some jurisdictions), the industry was prepared to complete wells with long sections of open hole behind the production casing. Most of the time, that worked just fine at the surface but who knows what was happening underground?

As far as the well owners were concerned, production could proceed smoothly, perhaps for years, but well safety depended only on the integrity of the casing. If the casing developed a leak or corroded through, all bets were off. And once the useful life of the well had expired and it was no longer generating enough money to pay for expensive abandonment processes, all too often the owners simply removed the surface equipment and walked away. Old oilfields are commonly peppered with these nightmare forgotten wells.

Drilling into the Reservoir

Drilling ahead – again

While the cement around the 9-5/8" casing is setting, preparations must be made for drilling into the reservoir. The reservoir rock, because we are reading a book about fracturing, will be a tight shale. A nice permeable sandstone or limestone would be easier to produce, but we are stuck with a nearly impermeable, fine-grained shale.

Shale should be easy to drill because it is fundamentally soft but at this depth and under this pressure, everything feels hard and resistant. Typically drilling progress might be 1 m/hr (3 ft/hr) so the 500 ft (150 m) section is going to take a week – of drilling time. Of course, you can't expect to drill continuously. Drill bits become worn and have to be pulled out and replaced, and pulling nearly 10,000 ft (3000 m) of heavy drill pipe out of the hole and running back in again takes lot of time.

We will be drilling with an 8-1/2" bit (216 mm) – a very skinny hole. If you recall we had the 9-5/8" casing laid out along one of your town streets – well, now you will have an even smaller diameter section at the end because we will run 7" (178 mm) casing into this hole.

Logging the Reservoir Section

Because we are in the most important section of the hole, we will want to learn absolutely as much as possible about the physical properties of the rock and the fluids it contains. We will need to run a full suite of geophysical tools into the hole and carefully measure and analyse everything we can.

This will take many hours and then the data goes off to the log analysts. They will be trying to distinguish any changes in rock type, any layering, any weak spots. They will want to know if there is any layering in the reservoir fluids, so production can be focused on hydrocarbon rich points rather than water rich ones. And they will want best estimates of porosity and permeability.

Interpretation of logging data is always more difficult for new and unknown reservoirs. Fields that are already in production will have the benefit of logging and production data from previous wells.

Liner or Full String

When the hole is ready, 7" casing is run either as a full string of casing right back to the surface, or as a liner. A liner is a length of 7" casing run on the end of drill pipe. It is just long enough to cover the open hole and up some distance into the 9-5/8" casing.

The decision on using a liner or full string is made by the drilling engineers and makes little difference to the well's integrity. Usually the cement operation for a liner is easier to control and that is the decider. If 7" casing is needed back to surface, the liner might be run first and then tied back to surface with a separate string of 7" casing. The cement operation for this second string is very much simpler because there is no open hole involved and no risk of fracturing during cementing. It is simply a matter of filling the annulus between the 9-5/8" and 7" casings.

Although running a liner is straightforward, cementing it is not. The annulus is tiny – no more than a finger width – and well conditions are difficult with high temperatures and pressures. The cement job will not be large, but it will be tricky.

However you decided to case this section, it is vital that you have a good and continuous cement seal around the pipe. You will be perforating the casing opposite the best bits of reservoir and you need to be confident you are communicating only with the formations you perforate. A bad cement job means you lose control of the reservoir. Gas can flow upwards between the casing and the hole wall. Unwanted water can also reach the perforations by flowing through poorly cemented stretches of annulus.

The last steps in well construction are to check the quality of the cementing very carefully, and to flush out all the drilling mud and replace it with brine containing chemical inhibitors against steel corrosion.

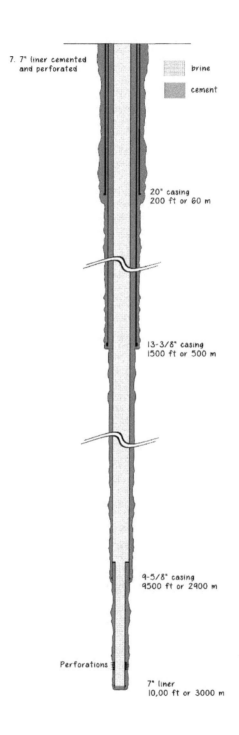

7. 7" liner cemented and perforated

brine

cement

20" casing
200 ft or 60 m

13-3/8" casing
1500 ft or 500 m

9-5/8" casing
9500 ft or 2900 m

Perforations

7" liner
10,00 ft or 3000 m

Perforating

Perforating the casing is a simple concept. A special carrier is lowered into the hole and positioned over the stretch you need to perforate. An electrical signal is sent and a series of small explosive charges detonate and punch holes through the casing and its supporting cement.

Unfortunately, because explosives are involved and we are ultimately talking about fracking, lazy members of the media have latched onto the idea that setting off explosives forms the fractures. Really – how dumb can you get? Perhaps they need copies of this book.

The job of a perforation is simply to open a pathway from the reservoir into the well. The explosives used are sophisticated shaped charges that send a jet of extremely hot gases through the casing and out into the formation beyond. In spite of years of experience and development, perforations are surprisingly small. Each charge punches a hole wide enough to stick your pinky in. The perforation tunnel penetrates perhaps a foot (30 cm) into the rock at most.

A single hole is restrictive and more holes make for easier flow into the well. A typical well might have 4 shots/ft (12/m) in the important places. As we will be fracturing, we will probably trade off penetration depth for a larger hole diameter because we need to pump large volumes of liquid through the perforations, and the fracture will be penetrating a long way into the formation anyway.

Completion

Once the well has been perforated and cleaned up, the drilling rig can be moved away. It is cheaper to move a small work-over rig onto location and the operator may choose to perforate with this smaller rig on site – it is all a matter of what works best for each particular field.

A conventional well is completed by running high quality steel tubing of perhaps 2-7/8" (73 mm) outside diameter down to just above the formation where a production packer is set. This is effectively a plug set in the casing which, when the tubing is not engaged, closes off the well. When the tubing is stabbed into it, production fluids can flow through the packer and tubing up to the surface – without coming into contact with the brine filling the space between the casing and the tubing.

If a well is going to be fracked, the production tubing and packer are run after the fracking operations.

So – where are we now?

As we are about to start fracking – the whole point of this book – this is a good point to stop and check over what we know. Let me make a check list, and you can go back and look at anything you have forgotten in the rush:

- We looked at permeable and impermeable rocks. A tight (impermeable) reservoir will need to be fracked to allow gas to be produced.
- The well will produce a mixture of gas, oil and water – collectively called produced fluids.
- We drill the well using drilling mud whose weight and viscosity is carefully adjusted to clean and support the well.
- The finished well is made up of a series of steel casings fitted together like a telescope
- Each casing size has been carefully cemented into place to bond it to the rock, and to isolate different layers within the rock.
- As we drill and cement, we have to be careful that the hydrostatic and dynamic pressure of any liquids in the well do not exceed the frac gradient – 0.7-0.75 psi/ft depth.
- Once the well has been cased to the full depth, the casing opposite the potential producing layers is perforated using explosive charges.

Are you happy with those ideas? It was a very quick run through a technical jungle and I hope all the numbers and all the different metric and traditional units did not leave you totally confused. I would say – don't worry too much about understanding the ideas in depth. Just look at the pictures and try to understand the implications of concepts like the frack gradient. If you can visualise the concepts even without fully understanding them, you will have enough background to understand what happens in the next step – the actual fracturing operation.

Before we leave this chapter, please have another look at the last picture (page 42) – of the well with all its casing in place. I wish I could have drawn it to scale so you would have a true mental picture of what a well looks like, but drawing a scale plan of a well is impossible. If we shuffle and squeeze so the length of the well fits on the page, the diameter of the hole and casing is far too small to draw. Less than a human hair in the lower sections. It just can't be done which is why I recommended you lay the well out along a street in town and think of it that way.

Now, let's get fracking...

The Anatomy of a Frack

We already know that if you increase the hydrostatic pressure acting against a reservoir formation, it will eventually break. That's just common sense. Stress anything enough, it fails.

Before we frack a well, we need to understand just what that failure will look like. So, let's consider a point in our reservoir 10,000 ft (3000 m) below ground and ask ourselves what stresses it is naturally under. The largest stress by far is the weight of all the rock above it; you would expect that.

What is less thought about – except by drilling engineers and geologists – are the horizontal stresses. These are built into the earth's crust and you can think of them as the horizontal squeezing stresses coming from the tectonic forces moving our continents very, very slowly sideways.

Engineers like to break the forces acting at a point down to three axial components. We already know the biggest one is the vertical weight of the rocks above. Let's simplify things a little and say that the horizontal forces can be thought of as acting in the north-south direction, and a second component in the east-west direction. Are these two horizontal components the same? Very rarely. In practice, one is always bigger than the other so we might have a big squeeze acting north-south and a smaller one acting east-west.

The question is – if I stress the rock so much that it begins to break what will the break look like? Think about that for a moment: you are physically breaking the rock and forcing it apart. How will it break? It will take the line of least resistance – that's the way life is. In this case, the fracture is not going to be a horizontal plane because that would mean lifting the entire weight of the overburden, and we know that is the maximum force acting on the reservoir rock.

The break will want to move the rock in the direction of the lowest acting force (east-west). If you are forcing liquid into the rock at a high enough pressure to break it, the rock will move equally to the east and west and create a vertical split running north-south. Please picture that.

Press your hands together and point your fingertips north. If you keep your fingertips and the heel of your palms together and open a space between your palms, you are moving your knuckles to the east and the west. The plane of the fracture is north-south, at right angles to the minimum stress.

Is the direction important? You bet. You want to know where the fracture is running so you know where to put your next well...

Natural fractures

You only have to look at a road cutting or quarry face or rocks exposed at the beach to know that they are layered and broken by natural fractures. Ideally, reservoir rocks are laid down in horizontal layers and develop fractures as temperature and the pressure due to depth turn the soft sediment into hard rock. In practice, reservoirs have been buried deeply for many millions of years and can appear quite jumbled up.

A word about natural fractures. These are created by the same horizontal forces we mentioned above. The rock is squeezed and, as soft sediment crystallises and later as hard sediment is cooked and squeezed at the same time, the orientation of crystals developing in the rock is determined by regional stresses as they grow.

Most rocks have a preferred direction to break in when you hit them with a hammer, and this is because all the little crystals are lined up in a particular direction.

Trouble is, regional stresses change with time. The rock may develop more than one set of natural fractures and you can commonly see at least two directional trends in sea-side cliffs.

How do natural fractures influence the hydraulic fractures we are making? Well, that depends. (You knew I was going to say that, didn't you?)

The physical properties of the rock are a major factor – and they are individual to each reservoir. The next most important factor is the viscosity and other properties of the liquid we are going to use for our frack job.

Speaking generally, a traditional viscous frack gel is more likely to produce a single planar frack. Modern slickwater frack fluids pumped at high rates in multiple stages are probably more influenced by natural fracture directions and give more irregular results.

The most important thing to remember about natural fractures is they don't actually exist. If they did, they would provide pathways for reservoir fluids to flow to the well, and we would not need to create artificial fractures. Reservoir rocks have a natural potential to fracture in a certain direction. When tectonic uplift and natural erosion exposes the reservoir rock at surface – tens of millions of years into the future – the unloading of overburden pressures will allow the rock to fracture in its preferred direction. Until then, all that exists is the preference, not the fracture.

Fracture Shape

Given that fracturing is essential to producing hydrocarbons from a tight well, engineers need to know what shape of fracture they can make. They need to know how long, how high and how wide it will be.

A single point fracture will start out like a penny, a semi-circular vertical fracture centered on the well-bore. As pumping continues, it will grow rapidly vertically and outwards. Vertical growth is limited by layering in the reservoir rock and most fractures reach their vertical limits quickly. The growth then continues outwards, away from the well bore.

The final shape of a classic fracture is two wings, stretching out on either side of the well bore. The engineers are most interested in how far the fracture reaches, and how wide it is.

The frack operators can control two main parameters – the viscosity of the fluid and the pump rate. Generally speaking, a very low viscosity fluid (known as slickwater) pumped at a high rate will tend to give long skinny fractures (provided it doesn't leak off). A more viscous fluid will tend to give fatter fractures that do not reach as far.

What stops a fracture growing? Fluid loss. As high pressure frack fluid is breaking open the reservoir rock, some of it leaks off into the rock itself. As the fracture gets longer and longer, its surface area is increasing and the area available for frack fluid to leak off eventually becomes so big that the pumps cannot keep up and fracture growth stops. This leak off is reduced by including a loss control agent in the fluid, but that is a two-edged sword. Coating the fracture face to reduce leak off may reduce its ability to produce hydrocarbons once the frack is in place. Gelling agents and fluid loss additives are helpful to make the fracture, but if they hang around afterwards, they tend to gum things up.

8. Schematic of a Developing Fracture

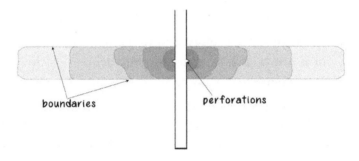

boundaries   perforations

The fracture is growing symetrically on either side of the well

Proppant

Once the fracture has reached its designed length and the pumps are shut down, it begins to close again. We had a nice pathway into the reservoir and, left to itself, the reservoir will close it off again very quickly as the pump pressure bleeds off. This is prevented by including proppant in the fluid – a granular material that props the frack open and stops it closing as pressure leaks off.

The selection of the correct proppant to use can be challenging, but the traditional best choice has been a high quality sand, with strictly controlled grain size and shape. This is adequate for most wells at shallower depths. If simple sand is used in deeper wells, it can be crushed as the fracture closes because of the forces involved. A variety of stronger materials have been evolved that are sand-shaped but much stronger. These are used in deeper wells but – they are very much more expensive.

Whichever proppant is chosen, the pumping rates and schedule are designed to carry the proppant as far as possible into the fracture, leaving a permeable route for fluid stretching perhaps hundreds of feet from the well. Proppants are selected to be stable and not degrade, so they are generally accepted as having no potential to pollute the environment.

Engineers try to get enough proppant into a fracture to hold it open but at the same time leave plenty of space between the grain for fluids to flow freely. A fracture packed with a layer of proppant several grains thick will be very much more permeable than the original reservoir rock, but is also at risk of collecting fine debris – such as fluid loss material or unbroken gel – and becoming plugged. Instead of completely filling the fracture with proppant, the engineers try to achieve what they call a partial mono-layer, which is science-speak for a fracture held open by only a patchy distribution of proppant grains. This is the ideal situation where fluids can flow freely but the proppant is less likely to be plugged by collecting fine particles.

A partial mono-layer is achieved by pumping slugs of low viscosity frack fluid with proppant, interspaced with slugs of simple frack fluid. Operations like this are referred to as slickwater fracks, as opposed to using more traditional high viscosity frack gels with high proppant concentrations.

Frack Fluids

You can frack a well with almost any fluid you can pump. In the early days of fracking, oil based liquids were popular – untreated crude oil thickened with corn starch (a first cousin to Napalm). You can fracture with CO_2 or nitrogen. Gelled acid fracturing is useful in reservoirs with a high limestone content, and foamed liquids can be used too.

But by far the most popular fluid for fracturing is water, thickened in various ways in order to carry the proppant, and that is what we will describe. The frack fluid has several jobs to do.

1. It has to initiate the fracture and allow it to grow.
2. It has to transport the proppant into the fracture.
3. It has to flow back leaving the reservoir (as far as possible) clean and undamaged.

Let's start with making the fracture. Initiating the fracture is easy enough – that could be done with plain water – but if you are going to create a fracture extending hundreds of feet into the reservoir, the first requirement is a liquid that is easy to pump. Pumping through the perforations and along the fracture at high rates will result in a lot of friction, increased pressure and a loss of efficiency. This is countered by including a **friction reducer** in the mix. These are typically manufactured acrylics or polyacrylics and are added at low concentrations, perhaps 0.2 % by volume. Along with all frack fluid components, these components need to be stable and not breakdown under the high reservoir pressures and temperatures they will experience.

The next job the frack fluid has to do is carry proppant. During a conventional frack job like the one we are working on at the moment, the frack fluid may have to carry considerable amounts of sand. At its climax, each US gallon of fluid may contain 12.5 lbs of proppant (that is, 1.5 kg of sand per litre). The frack fluid must be carefully engineered to do this job, by including a gelling agent in the mix. By far the majority of fracturing gels are based on guar gum, derived from the guar bean. It is a natural polymer, cheap and easy to produce and biodegradable i.e. easy to dispose of.

Another useful feature is that it is easily cross-linked. The original polymer molecules take the form of chains of atoms; if a **cross linker** is introduced into the gel, these chains tend to join together side by side. The result is still a liquid, but one with very strange properties. A favourite demonstration for frack experts describing their trade is to take a beaker of cross-linked gel and start to pour it out. When nearly half has left the beaker, the beaker is tipped back to upright, and the gel will flick back into it without a drop spilled. And the liquid is still pumpable.

You can imagine that a frack fluid of this type is very capable of carrying a lot of proppant into the fracture.

The final job a frack fluid has to perform is, basically, to disappear. It has done its job of opening the frack and placing the proppant, now we want it to lose its viscosity and flow back to the well leaving no trace behind. In practice, there is always some debris left - improperly hydrated gel, fluid loss material or just general dirt. For best results, the spent gel must be actively broken and a lot of the breakdown is achieved by the downhole temperature alone. The gelling agents will breakdown if over-heated. To give a more complete and faster break down, the frack fluid will include a **breaker**, usually an oxidising agent. The better the breakdown, the less likely we are to plug the packed fracture with debris.

Frack Fluid Components One by One

Water

The largest part of a conventional frack fluid is water. Not just any water but clean, filtered fresh water with a minimum of dissolved salts. In some areas, there may be trouble finding enough good water at a sensible price. Water shortage can be a real issue on the new shale gas fields, especially if unconventional frack techniques are used. And if it can be found, the thousands of truck trips needed to bring it to the well can have a major impact on local roads.

Friction reducers

As mentioned above, these are mostly acrylamides and polyacrylamides, sometimes delivered as powder but more often as emulsified liquids to aid in mixing with water.

Perhaps now is a good time to mention how most fracking additives are developed. Very, very few oilfield chemicals are developed by the oil industry. In most cases, a need is defined and a search begun for an existing industrial product that will fit the bill. This is a very cheap and efficient way of getting what is needed, and the imported chemicals come with all the regulatory approvals already in place. Using well known chemicals makes life much easier for the fracking companies.

This process is normal for friction reducers and other additives.

Gelling Agents

A good gelling agent enables frack fluid to carry proppant, reduces the tendency to leak-off into the reservoir, and reduces friction while pumping. The go-to material is guar gum and various derivatives, because it is a cheap, accessible material widely used in food products.

Guar comes from a bean grown in tropical countries that is simply dried, cleaned and ground up. When carefully mixed with water it forms a long chain polymer gel that can be broken down to a very clean residue – very important if a fracture is not to be plugged up with debris after the job.

One of the attractive features of guar based fluids is that they can be easily cross-linked to vastly improve their carrying capacity. Cross-linking means the long polymer chains are encouraged to join in parallel with each other by adding a cross-linking agent as a catalyst.

There are other gelling agents based on surfactants or emulsifiers, but these are very much less common. As they tend to be more expensive, they are only used where reservoir conditions demand them

Cross-linkers

The cross-linking of frack gels is highly sensitive to pH conditions (that is, the acid-alkali balance). When the pH is right, a very small amount of cross-linking agent can turn hundreds of barrels of frack gel from a simple viscous liquid into one that can carry a great deal of proppant without difficulty.

The commonest cross-linking agents are borate based. They cover most bottomhole conditions, and allow the gel to be easily broken and produced back. Borate salts are common in the environment and that makes disposal easier.

For higher temperatures, titanium and/or zirconium cross-linkers can be useful.

Breakers

Breakers are included in the frack gel to accelerate its breakdown once pumping stops and allow the gel to flow back to the well bore easily during clean up. An effective break of the gel is important as partially broken gel can leave debris that becomes 'cooked' at bottomhole temperatures and hinder flow through the propped fracture.

Most common breakers are oxidising agents, usually ammonium and sodium persulphates or calcium and magnesium peroxides. Ideally they should lie inert until the job is finished and then come into action during the clean up and some encapsulated breakers have been developed to achieve this.

Enzyme based breaking compounds have also been used, and the fracture can be over-flushed with hydrochloric acid to break residual gel. A note here on hydrochloric acid use in the oilfield. There is no doubt that hydrochloric acid is a very dangerous chemical and you certainly do not want to spill it on yourself or get it into your eyes. However, it is not persistent. Spill some on the ground and within a minute or two it will probably have turned itself into harmless chlorides and water. Hydrochloric acid is a mainstay of the oil industry and widely used for cleaning up wells and stimulating production. Handled with care, it is not a worry. What can be a worry are the corrosion inhibitors commonly pumped with it to prevent damage to downhole tubing and casing.

Stabilisers

Commonly methanol or sodium thiosulphate, these are added to stabilise the gel properties at higher temperatures.

Buffers

Chemical buffers are used to control the pH of the basic gel and are vital to cross-linking. The most common for increasing pH are sodium carbonate, bicarbonate or hydroxide. Lowering pH is done with simple acids.

Surfactants

As the well is being cleaned after fracturing, a surfactant is a useful cleaning aid – think of them as being like shampoo or kitchen washing up liquid. In fact, such simple surfactants are acceptable in some wells, but more complex ones are more commonly used because it is important to leave the reservoir and fracture surfaces water-wet and not coated in oil. Mutual solvent surfactants are used, mostly glycol monobutyl ethers that are effective and easily disposed of.

Biocides

Frack fluids are basically organic and, left to themselves, will grow micro-organisms capable of breaking them down under surface conditions. To stabilise the fluids, a biocide is included in the mix. By definition, the biocide is the component of a frack fluid most potentially hazardous to the environment and must be handled with care both before the job and in the flow back phase.

Frack Additives and Secrecy

In the early days of fracturing, the world's attitude to environmental stewardship was much more relaxed than it is now. Regulatory authorities really did not care too much what chemicals escaped into the environment and so not many questions were asked.

Or answered, come to that. The people who were developing the technology of frack fluids were the same service companies that performed the frack jobs. They had every interest in preventing their competitors learning exactly what was in their frack fluids.

They achieved this secrecy by labelling all the components with brand names or product codes, such as FR-56 or J347. If you asked what exactly was behind the product code, you were told that was proprietary information. The oil industry were not the only ones playing this game with the public. A few years ago most of the products in the cleaning aisle of your supermarket were equally secretive.

That attitude is understandable but ultimately self-harming. The sophistication of the general public grew over the years, and they also became much less likely to trust what they were told. They became convinced that behind the product codes were lurking some very poisonous chemicals. Both the oil companies and the fracking companies tried to keep the books closed but ultimately, the regulatory authorities had to be told exactly what was being pumped.

The same authorities needed to know exactly what hazards were in water produced back to surface after fracturing

No-one wants to see dangerous chemicals in a frack fluid. After a frack operation, the spent frack fluid is produced back and must be disposed of by the well owner. If a really damaging chemical, such as toluene, is included in the frack fluid, the well owner will be faced with the problem of disposing of large volumes of contaminated flowback. Self-interest means the well owners pressure the fracking service companies to be as environmentally friendly as possible.

The ultimate irony is that most of what is included in a frack fluid is more or less harmless, but the general public finds that hard to believe and who can blame them? Of course, if you have been told for decades that the real identity of an iron control agent is highly confidential, you are going to be suspicious of sudden openness. Now we are told that the highly confidential chemical is actually the widely used food additive the tetra-sodium salt of ethylene di-amine tetra-acetic acid (EDTA), we are entitled to ask – why all the secrecy?

In fact, nowadays the secrecy has largely disappeared. If you look at http://www.halliburton.com/public/projects/pubsdata/Hydraulic_F racturing/fluids_disclosure.html or http://fracfocus.org/ you can see in detail the chemicals that are in common use. The fracking companies and their employers the oil companies now take the attitude that is impossible to hide the make-up of frack fluids from the government and so they had better make the best of the situation and tell the truth. That does not make the few dangerous chemicals in common use any less dangerous, but at least now we all know what those dangers are, and how to counter them.

Let's Get Fracking

Now we have our hole drilled and cased; the reservoir has been logged and perforations made just where we want them; we have decided on the size and type of fracture we want to create and, as a final step the suspiciously friendly engineers from all the service companies have dropped by and left their fracturing proposals, we are ready to start.

It is a commonplace that operations like fracking are 90% complete before the first pump is fired up – meaning that there is a great deal of logistical and preparatory work to be done. Over the days before the frack, the service company doing the job will be placing equipment and supplies at the well site.

Equipment

Tanks

The most obvious equipment will be the tanks for holding frack fluid. In North America these are normally 500 bbl monsters (30,000 litres) with road wheels built into their back end, to save using a crane to unload them on site. The number of tanks depends on the size of the operation – it might be two, it might be twenty or more.

The tanks are carefully lined up so they can all be joined to a common manifold. The manifold and connecting hoses will all be of large diameter so fluids can be sucked through them at low pressure.

Proppant Storage and Handling

During the job, proppant will be mixed with the gel, so it must be stored on site in special silos or tankers that are capable of delivering a controlled flow of proppant at high rate. This can be tricky if more than one silo will be needed as all the proppant storage has to be sited around the mixing equipment and things can get a little crowded.

Mixing

At the heart of the operation will be the blender, where liquid and proppant are mixed together and pumped on to the high pressure pumping units. This is a critical piece of equipment and needs a very skilled operator. The blender delivers its gel through a low pressure manifold to all the pump units.

Pump Units

For all but the tiniest fracks, multiple pump units will be needed. They are the queen of the show, earth-shaking monsters capable of pumping at very high rates AND very high pressures. The experience of seeing ten or twenty of them lined up together and working hard to force large volumes of frack fluid into the reservoir is very impressive. If you get the chance to see them in action, don't miss it. It will stay with you forever, good or bad.

They could be pumping at pressures up to 20,000 psi (140,000 kpa) – very high and very dangerous.

Treating Iron

Connecting the pump units to the well head is specially designed treating iron. This connection must be capable of handling a very abrasive mixture of proppant and gel at high pressure, but it must be physically flexible and capable of being strung together and later removed quickly and easily.

Included in the high pressure lines will be valves, and check valves to prevent flowback in event of a breakage, flexible loops and elbows to absorb vibration, and a complex connection to the well tubing. The well head assembly will include pressure and rate monitors.

Control and Monitoring

During a complex operation, the person in charge – the frackmaster – needs instant communication with all the equipment and its operators. There will usually be a control van off to one side where he presides over his team, keeping the operation running to plan and ready to shut it down instantly in the event of a problem.

Any significant frack operations will also have a monitoring van where operational data is recorded and some real time analysis done. In contrast to the noise and thunder outside, this van is usually quiet and calm. And has the best coffee.

Site Preparations

Once the drilling rig has moved off, the area around the well head is left flat and clear, ready for the frack equipment. After the tanks have been sited, there will be a stream of water trucks filling them. On a big operation, this can take some time, and put a strain on the local roads.

Once the tanks are filled, the blender and hoses are connected up and the gelling agent can be mixed into the water. This is usually done on site, and at the latest possible time to avoid degradation in the gel.

As the tanks of water are mixed, the quality of the gel is checked. The viscosity is important but the pH and temperature are also critical. Any shortfall in quality has to be corrected and in countries with cold winters, it may be necessary to bring in heating trucks to bring up the temperature.

While this is going on, a hundred and one other details are being attended to. A big frack job will require a big crew, and they must be fed and watered while they work. Environmental protections must be checked, and may be inspected by the regulatory authorities. There may be landowners to deal with, and local media are likely to show an interest. Added to that, casual sightseers may wander too close and have to be steered away.

The Big Day

When everything is ready, the site is handed over to the frackmaster and his crew. They will generally start early in the day, testing their equipment. All the pumps and treating lines will be pressure tested to pressures much higher than expected during the job and any faulty equipment replaced and re-tested.

No possibility of a leak can be tolerated because a leak of high pressure, sand laden gel can quickly cut through heavy steel and shut down the job before it is completed. Not to mention the safety and environmental implications of a leak.

The first step of the frack job may be to test injection and frack pressures by pumping a little gel into the formation, and the job can begin.

A big frack job is very impressive. During the first minutes, the blender equipment and auxiliary pumps are fired up, and then joined by the big high pressure pumps. The flappers on the pump unit exhausts lift and their great engines begin to rumble. If there are turbine powered pumps on site, their big outlets doors swing upwards and the thunder of the diesels is joined by the deafening whine of helicopter turbines. The air above the pump units becomes hazy with heat and fumes and the ground vibrates.

Inside the frack bus where the pump operators sit with their consoles, everyone is concentrating on the performance of their particular unit. Behind them, the frackmaster monitors flow rates and pressures, and talks the job through its planned phases.

The first step will be to pump a gelled pad – a volume of gelled fluid without proppant that has the job of filling the tubing and initiating the fracture. Then proppant is delivered to the blenders, starting at low concentrations and stepping up as the job progresses.

Outside, the site is empty. In particular, no-one gets anywhere near the pumps or treating iron. Any mechanical failure here could kill in an instant. Despite the power of all the pumps forcing fluid into the well, there is little vibration around the well. Without the noise, you might not understand anything was happening.

The last phase of the operation – once the fracture has been formed and filled with proppant – is to pump another pad of gelled fluid. After the fracture has closed, the well will flow back and the gel will be needed to carry any stray proppant out of the well.

After the Party

Once the fracturing operation has been completed, the well is carefully produced back to remove as much of the spent frack fluid as possible. Frequently the well will not flow without assistance and a coiled tubing unit may be brought in to pump nitrogen into the bottom of the well to assist with clean up.

The fluid coming back to surface will contain the remnants of the additives used in the gel, plus reservoir hydrocarbons. Normal procedure is to collect this in a lined pond on site for later treatment and removal. It will not be clean enough to release into the environment.

After a day or two, the well site will be clear of equipment and, if the flowback period has been successful, the well will be ready to deliver its hydrocarbons into a pipeline. Very soon, the well will be no more than a lonely Christmas tree in a small fenced enclosure, silently producing useful (and saleable) hydrocarbons day in, day out.

Horizontal Wells

You have now drilled and fracked your first well, a conventional vertical well with a fracture reaching out into the formation. This well works by making it very much easier for hydrocarbons in a tight formation to reach the well bore. The next technical development has to do with the length of producing well in the reservoir. A vertical well penetrating a 100 ft thick reservoir draws its production from only 100 ft of rock. If you could drill down to the reservoir and then turn the drill bit to go sideways, you would be able to access many hundreds of feet of rock.

Ever since rotary drilling for hydrocarbon was in its infancy, there has been a need to drill deviated wells. If there was a drilling problem and a bottomhole assembly became stuck in the well, you had two choices. Fish it out, or set a cement plug over it and drill a deviated hole off to one side to avoid the obstacle.

The ability to drill wells at an angle was initially mysterious and thirty years ago the directional drilling specialist was treated as something between a High Priest and a Grand Wizard.

Times have changed, especially with the importance of off-shore oil fields. Now oil companies routinely drill a network of wells from each of their platforms, and they need to know just where those wells are going.

The technology of deviated drilling has leapt ahead and now the drill bit can be tracked in real time and steered to its target. Drilling under towns and cities has become routine. In some places, offshore oil can be reached from land by horizontal wells reaching many thousands of feet under the sea.

Tight Formations and Horizontal Wells

If you recall, we fracked our vertical well to reach out into the reservoir. Could we achieve the same results by drilling horizontally? Yes, more or less. It is common sense that a long horizontal hole will give access to more of the reservoir than a small diameter vertical one.

The development of horizontal drilling came at a time when the United States was becoming more and more dependent on imported oil. The demand for more home produced hydrocarbons plus a steadily climbing price for them stimulated oil field engineers to look more closely at the techniques that were already on hand, and apply them to reservoirs that were slow becoming more attractive as prices climbed.

The service companies developed new Measurement While Drilling (MWD) techniques and equipment and soon horizontal wells could reach far out into reservoirs. Fields that had been uneconomic to produce became mainstay producers.

Drilling a Horizontal Well

Except for the reservoir section of a well, there is little difference between a horizontal and a conventional well. If we take the well we have just drilled and fracked, there would be no difference until we got well down the 12-1/4" section to, say, 8-9,000 ft (2,500 – 2,750 m). Then we would have to decide how we are going to start deviating the well to horizontal.

If the formations allow, we might decide to set the 9-5/8" production casing at, say 8,500 ft (2,600 m). Once the production casing is cemented and secure, we can drill ahead a vertical hole through the reservoir and log it. This allows us to select the best position for our horizontal hole – the sweet spot where production will be maximised. We plug back the vertical hole with cement and wait for it to set rock hard.

While this has been happening, the specialists are setting up their monitoring equipment. Their aim is to steer the drill bit from vertical to horizontal and keep it at just the right level.

There are a variety of measuring tools that can be run behind the bit, not only to give the real time position of the bit but also to measure formation properties as it drills.

The heart of the downhole assembly is the mud turbine. It is not practical to control a rotating drill string as the angle of the hole builds to horizontal. Instead, the bit is mounted on a mud turbine motor that is turned by the flow of mud being pumped through it. This means drilling is not done by rotating the drill string as normal. Instead the string stays steady, sliding down the hole as the mud turbine drills deeper.

Using the mud motor to turn the bit and by listening to the complex tools immediately behind it, the hole can literally be steered into the position we need.

The experience and technology developed over the last 25 years means that this sort of drilling is now very common. Horizontal wells can be drilled easily and economically, opening up new classes of reservoir for the first time.

Why Drill Horizontally?

If we set aside the need to drill horizontal wells under built up areas, or out under the sea, the biggest attraction of a horizontal well is the way it reaches out into the reservoir.

Let's imagine we are looking at a conventional well from above and can see the way hydrocarbons flow into the well-bore. Looking at Fig 9, you can see the flow lines forming a pattern like a dart board, moving from an uncrowded outer area into a focus where they become very crowded indeed.

9. Radial flow to a Vertical Well
The flow lines crowd together as they reach the well - meaning the flow of hydrcarbons will be very restricted

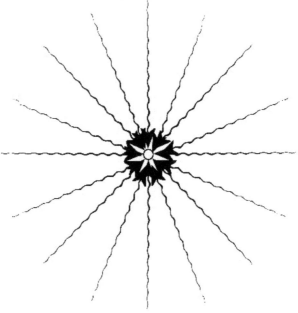

In a very permeable reservoir, this may not be a problem. Fluid can flow through a permeable reservoir quickly and congestion near the well may not be important. In a tight reservoir, things are different. A tight reservoir simply will not allow fluids to move quickly and all the fluid in outlying parts of the reservoir is held up by the bottleneck near the well bore. Even with the benefit of deep perforations, the fluid has to squeeze into a circle about a foot (300 mm) in diameter, and that is just too hard.

In the old days, tight reservoirs were simply unproduceable, and were ignored. The hydrocarbons were in place but took so long to get to the well bore that it was not worth drilling a well into the reservoir.

The simple solution is to frack the well, as we did in our example. Now instead of trying to squeeze into a one foot circle, fluid can access the well by flowing into the frack plane, and that might be hundreds of feet long – Fig 10. Drops of fluid are still moving very slowly through the tight rock but, because of the length of the fracture, so many more are able to move in at once that the well becomes an economic proposition.

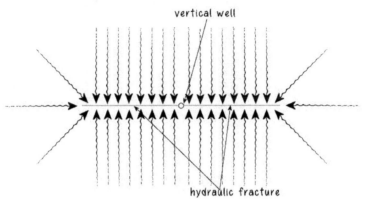

10. Flow to a Fractured Well

Dispersed flowlines indicating much easier access to the well for hydrocarbons.

And horizontal wells? Well, the same reasoning applies, except that the length of well bore in the reservoir can be several times what can be achieved by fracking a vertical well. More penetration of the reservoir means more production. Just drilling a horizontal well instead of a vertical one helps flow from the reservoir, but if the horizontal well is also fracked, even a tight reservoir can be drained very effectively – Fig 11.

II. Flow to a Fractured Horizontal Well

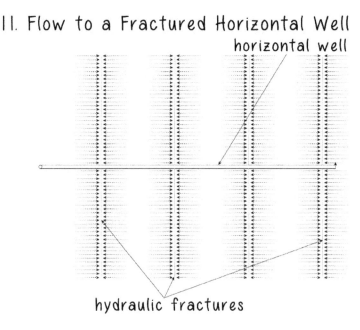

horizontal well

hydraulic fractures

There are reservoirs that are too tight for economic conventional fracking but come into the productive zone if horizontally drilled. There are many, many more that need both horizontal drilling <u>and</u> fracturing to be economically successful.

Fracturing a Horizontal Well

We have already fractured a vertical well, where the priority was to create an easy path for hydrocarbons to flow to the well from as far away as we could manage. We created a single fracture with two wings stretching hundreds of feet either side of the well bore.

Fracturing a horizontal well is just the same, with the exception that we don't need to restrict ourselves to just one fracture.

The first step in designing a horizontal well for fracturing is to determine the direction the reservoir naturally wants to break in. This is found from seismic studies, and especially micro-seismic studies of previous frack jobs. Micro-seismic gives us a very good picture of the orientation and length of induced fractures.

Let's suppose the reservoir naturally fractures in a north-south direction. To get the best out of it, we will drill our horizontal well in an east-west direction, cutting straight across the natural fracture trend.

Once the horizontal section has been drilled, the next step is to case it, and cement the casing. This cementing is difficult because the casing wants to lie on the bottom side of the hole and we need lots of strong centralisers to lift it clear. Of course, lots of centralisers make running the casing difficult and the drillers will complain bitterly. Centralisers have been known to accidentally disappear into the mud pit...

The quality of the cement job is important to the success of the frack jobs. We need to be sure that when we frack in one position, our frack fluid is not diverted along outside the casing to other parts of the reservoir.

Once the casing is cemented and tested, we are ready to frack. A perforating gun is run to the far end of the hole and several perforations made. Perforating debris is cleaned out and tubing run into the hole with a packer on the end. The packer is set just before the perforations, and a frack job carried out.

The packer is pulled from the hole and a drillable plug run in and set before the perforations, leaving just unperforated casing exposed. The procedure for the first frack job is now repeated – perforate, clean, run tubing, frack, run a drillable plug. And we can keep doing this along the horizontal length of the well. The size of the frack jobs and their spacing depends entirely on the reservoir properties.

As far as any passer-by is concerned, there is no difference on the surface between a vertical and a horizontal well. They are completed and produced in exactly the same way, they have the same well head equipment and they take the same amount of space. The real difference is that horizontal wells can make money in places that vertical wells can't.

Coal Seam Gas

Coal was once a very valuable mineral resource and the prime source of energy for the Industrial Revolution. It is much less valuable now. If you compare it with oil or natural gas, it is heavy and difficult to transport. It is also very dirty to burn and today that is a serious problem. Coal's significance on the energy market is declining year by year.

No-one is interested in mining it unless it is easy to get at, and in an area with compliant environmental regulations.

However, that is not the end of the story. Coal beds are always associated with methane gas. It used to be known as firedamp and was responsible for taking many miners' lives a hundred years ago. Any naked light could start a lethal explosion. You may recall stories of miners taking canaries down below to warn of gas – that is the gas we are talking about.

It may have been a deadly nuisance to miners, but it has spawned a whole new sector of the hydrocarbon industry. If a well is drilled into a coal seam, it can be de-gassed in advance of mining. And in addition, there are many coal seams that are too deep or too thin to mine but still hold enough gas to be interesting.

Drilling for Coal Seam Gas

It should be easy. The coal seams we are interested in are typically between 300 ft and 5000 ft deep (100m – 1500m), so only light drilling equipment is needed. The reservoir pressure is low compared to shale gas and that makes it easier to handle. And coal itself is easy to drill.

The fundamentals of drilling a coal seam gas well are exactly the same as a deeper shale gas well but the difference, and the difficulties, lie in its shallow depth.

If you recall laying out the casing strings for a deep well along your town streets, think how different things would look if your main casing string only extended a couple of blocks.

The coal seam reservoir can be shallow and therefore very much nearer to drinking water aquifers. Any problems may not be happening thousands of feet deep and insulated from the surface by thousands of feet of cement seal. They may be happening much nearer home.

That is the trade-off. The wells are easy to construct, repair and maintain, but small problems can become big ones very quickly.

As always, we do need to appreciate the scale of the wells and the aquifers through which they pass. Wells completed in the deeper coal seams at 5000 ft (1500m) are way below normal water extraction wells and achieving isolation from aquifers is no harder than for shale gas wells. Shallower coal seams might be immediately below a working aquifer. Sealing the well bore off from the aquifer is vital, and fracking might not be possible if the fracture height might break into the aquifer. (If there is a risk of this last happening, the operator will not wish to frack at all. He does not want to be stuck with producing even more water along with his gas.)

There is a significant difference between coal gas and shale gas, apart from the scale of the wells. Gas in shale is more or less free to move about, but the gas in a coal seam is adsorbed onto the coal itself. It only detaches itself when the surrounding pressure is reduced – that is, once you start removing water from the coal seam. As the water is pumped out and the reservoir pressure is reduced, methane is slowly released.

The water problem in a conventional gas well is to dispose of water that comes along with the gas. In coal seam wells, you don't get any significant gas unless you pump water from the reservoir. Comparatively, produced water is more of a problem with coal seam gas.

Geomicrobial Prospecting

This may seem a strange place to include a chapter on a hydrocarbon prospecting techniques, but bear with me.

Ordinary soils around the world are home to a host of micro-fauna and flora, little bugs busily living their lives in an ecosystem most of us never see or think about. I am told it is a whole new world and that microbiologists still have much to learn about it.

One of the things we have learned over the last half century is that certain organisms like hydrocarbons. Conveniently, there are two common classes of hydrocarbon-loving microbe, one of which really enjoys methane in its diet and another that will only look at heavier gases like ethane, propane and butane. If you take shallow soil samples from the fields and forests over a gas reservoir, you will find increased levels of the methane-loving bacteria. Over an oil reservoir, you will find the other sort. And where there are no hydrocarbons below, there will be virtually none of these useful bugs.

The technique is not perfect and works more or less well in different places. It can be extremely useful in extending existing fields as you can use the bacterial pattern over the known oil or gas field to compare with neighbouring areas.

The main attractions of the technique are that it is quick and very cheap. No specialised field equipment is needed. The field crews carry a hand-held GPS and a trowel or hand auger. Samples are collected in a grid pattern all over the search area and sent back to a central laboratory where the microbes are cultured and counted. A map of hot spots and dry areas can be on the oil company's desk in very few weeks.

The results are not guaranteed, especially in new ground. No-one would drill a well based solely on the geomicrobial results, but they are quite likely to examine the hotspots more closely, usually with a seismic survey.

What do Geomicrobial Results mean for us?

All through this guide, I have been emphasising questions of scale, and this is another example. A good gas or oil reservoir is overlain by impermeable bedrock. The hydrocarbons cannot escape unless we drill a hole into the reservoir and allow them a route to the surface. Except, that is only *relatively* true. If you are a geologist and think on a geological time scale, the idea of an impermeable reservoir seal gets very problematic.

Taken over tens of millions of years, no rock is impermeable. Not even glass. Gas is slowly seeping through the most impermeable of rocks. Very, very slowly, in the case of good reservoir seals. In other geological contexts, perfectly good hydrocarbon bodies have leaked all their volatile fluids to surface over hundreds of thousands or millions of years, leaving tarry rocks that are useless as energy sources.

And so? Why should we worry? We must be aware of the slow leakage that nourishes the microbes because it means that testing over a reservoir will show traces of hydrocarbons, including the nasty BTEX chemicals, in the soil and water even before wells are drilled. They are naturally present and we live with them every day.

What has this got to do with Fracking?

Nothing – directly. It does mean that the soils over a reservoir, even a deep reservoir, may naturally show traces of hydrocarbons. If we are to be sure of the impact of drilling a well and producing hydrocarbons, we need to be aware of these background levels before we start.

What could possibly go Wrong?

Drilling for hydrocarbons and producing them is a human activity, and we all know what that means. Things will go wrong. Things always go wrong, no matter how hard we try.

Before we start, it is important to distinguish between problems caused by accidents and ones we cause ourselves. If society allows a factory to pour its waste into a river or release it into the atmosphere, that is a self-inflicted wound. It is a planned and approved activity, not at all the same as an accident causing a diesel transport tanker to turn over and leak, or gas escaping from a badly constructed well.

We live with the risk of things going horribly wrong every day. Human beings are remarkable for the way they automatically assess dangers and risks, and then put them to the back of their minds as they go about their daily business. We jump into our cars and confidently drive on two-way roads, only feet away from similar cars approaching us with a combined speed of 100 mph (160 kph), and when did you last worry about that?

Two things contribute to our complacency. We know from experience that the chance of an unprovoked head-on collision, on dry roads in clear weather, is very small indeed. And we also know we are wearing our seat belt, we have airbags, and the car designers have done their best to make collisions survivable.

Even safer than driving is taking an airplane. We know that when things occasionally go wrong, a plane crash is nearly always a complete disaster, but we still fly. We have confidence in the planes, in the people who operate them, and the regulatory framework that controls every aspect of our flight.

So how does our fracked gas well differ from the aviation experience? Can we be as complacent about that?

Sorry, but no. There are host of technical things that could go wrong and I'm going to discuss some of them below. Before I start on the technical details, let's look at two general points. Two potential causes of disastrous results, and I expect they will surprise you.

A Fall in the Gas Price

The oil and gas industry is famous for its ability to quickly mobilise money and resources to drill wells and make their owners very, very rich in no time at all. America and its free enterprise philosophy excels at getting the job done and making a good profit. Nothing is as profitable as a successful oil well, and nothing returns your investment so promptly – usually a matter of months.

Of course there are good wells, and not so good wells – meaning wells that are cheap to produce from and wells where the cost of production takes a big bite out of potential profits.

Wells that need to be fracked are, by definition, in the not-so-good class. Their tight reservoirs mean that even after fracturing, production is low. It is also likely to decline more rapidly and typically production will decline two-thirds in the first year or two. That's life. We are talking about an extractive industry in difficult reservoirs.

This natural peak and decline in production can tempt operators into Ponzi scheme thinking. Flush with cash from initial production, they might choose to drill and drill but, ten years down the road, they will be sitting on a portfolio of wells producing only a fraction of their initial flows. These wells are not deadbeat assets. Far from it. They will still be producing good gas, but not in large quantities. It might be time to sell out to a smaller operator and move on.

Now let's introduce the killer variable – the gas price. When it is high, everyone is happy. Even producers with difficult wells. When it is low, those difficult wells do not even get drilled.

In a developing field, price fluctuations are less important, because the first production flush has probably covered costs anyway. For mature fields, price fluctuations are a different story. It does not cost much to run an existing well location. Most of the operating cost of a mature field goes into gathering, processing and compressing gas for sale. The individual wells simply need monitoring as they produce low volumes of gas in a long slow decline.

The trouble comes near the end of a well's working life. Perhaps it stops producing enough gas along with the inevitable water. Perhaps it develops a plugged formation or even a downhole leak. At this point the operator must decide if it is worth repairing or re-stimulating, or whether it is time to call it a day and abandon the well.

This is when things get challenging. Abandoning a well safely costs money, and big, responsible producing companies budget for that. However, if a field in terminal decline is in the hands of a cowboy operator, there may be no money available to abandon non-productive wells properly. Who picks up the tab if they go bankrupt? You may say that, no matter what, abandoning a well properly is the responsibility of the owner, and that is true. However, people are people and you cannot put much trust in shady business deals. Society needs to appoint effective regulators to make sure businesses are only allowed to own and operate wells if they can guarantee their safe abandonment when their productive life ends.

A significant drop in gas prices with little hope of an immediate recovery can put a lot of fields into trouble, and the temptation will be to save money and cut corners. The regulatory framework must be able to handle this.

The Good Ole Boys

In terms of the number of wells drilled every year and the dynamism of its oil field industries, the United States is unique. It leads the way in just about every aspect of the international oil business.

Unfortunately, the development of the drilling and production side of the industry has been hobbled again and again by a peculiarly American phenomenon – the Good Ole Boys.

This is a touchy area because so much of what we know today has come from hard working and intelligent workers from the rigs and oil fields. Most of these men (they are nearly all men, I'm afraid) represent a depth of skill, knowledge and experience that is literally invaluable. They are listened to within the industry, and held in respect.

Unfortunately, some of them are not so valuable. The worst of them place their own judgement and experience far above everyone else, and they tend to have little respect for higher education. Am I being too harsh? Well, in the 80's, I worked on two delta rigs in southeast Asia with managers (tool pushers) from Louisiana. They were running multi-million dollar operations but could hardly read and write, and had to pass on even the simplest of calculations to the mud engineer (who checked his own calculations). These men were both in their thirties at the time, very capable and good managers. Where did they go when their time overseas had finished? I wonder.

There are still men in positions of importance within the American oil industry who are not well equipped for the decisions they have to make. If you want a concrete example, look at the official reports of the disastrous 2010 Deepwater Horizon blowout. This was a technically demanding well, drilled to great depth and in deep water, but the sequence of seat-of-the-pants decisions made both on the rig and in the onshore HQ show that, yes, the Good Ole Boys are still there.

They are the oil industry's worst enemy and a real stumbling block to progress.

Potential Problems with Fracked Wells

We already know that getting a well ready for production is a very complex process and it will inevitably have impacts on the environment and people around it. Lots of things could go wrong, some important and long lasting, some of little concern at all. Perhaps the best way to look at them is to list the common concerns against hydraulic fracturing.

Fracking Pollutes Groundwater

This is the loudest complaint against fracking and a very serious concern for people living near the wells. Let's separate it into two – pollution caused by the fracking itself, and pollution caused by the well construction.

Fracking pollution flowing back to surface

If you remember how we laid out the casing of our example well along the streets of your town, you will quickly see how unlikely this is. Even the biggest conventional fractures only reach out hundreds of feet horizontally in a reservoir that is thousands of feet deep.

You may have a question about the limits to the vertical growth of our fracture because we cannot actually see what happens. Scientists can 'look' at fracture height development on real jobs using microseismic techniques and extensive studies confirm the theoretical prediction – fractures do not grow vertically far out of the predicted zone. There may be some penetration along existing fault lines but this is rare and anyway does not extend far out of zone.

In other words, conventional hydraulic fractures cannot and do not extend the thousands of feet upwards that would be necessary to pollute fresh water aquifers.

Coal seam gas wells are a possible exception, if they are very shallow. A shallow well directly below an existing aquifer may be too dangerous to frack.

When thinking about gas flowing to surface, we need to think how it happens. The reservoir and surrounding rocks hold gas under pressure. We make the gas flow by effectively reducing the pressure in the well bore. Fluids always flow down pressure gradients so both gas and liquids will be heading for the well bore only.

Leaks from the Well Bore

By drilling a hole, we provide a possible route for fluids to move out of the reservoir towards the surface, and for water to flow between the layers of rock we have penetrated. We use drilling mud to keep a positive pressure against the rocks and stop this happening during drilling. Once drilling has been completed, the well bore is sealed off by casing enclosed in a cement sheath.

So – how can fluids move vertically in the well bore? For that to happen, either the cement sealing the casing into the well was not placed properly, or it is not there at all.

Cementing casing is a difficult operation and there is always the chance it will be only partially successful. Wireline tools exist to check the success of cement jobs so the well owner should know if there is a potential problem or not. Some jurisdictions insist on bond logging, others leave it to the well owner. Unbelievably, the operator of the Deepwater Horizon well (through his Good Ole Boys) decided to economise and not bother with a bond log…

Casing may be well cemented and prevent vertical flow, or it may be poorly bonded to the rock in places. This does not mean vertical flow can happen freely. If you return again to the well casing laid out along your street and imagine water wants to flow up the outside of it, you soon realise that, no matter how long it is, all that is needed to stop the flow is a relatively short section of good cement. A few feet of successful cement now and again will prevent the flow but – and this is important in the long term – any uncemented sections of casing will be exposed to corrosion from formation fluids. Some of these fluids can be aggressive brines and a casing failure through corrosion will simply be a matter of time.

So – a simple question: can fracturing pollute drinking water aquifers? Simple answer - no, and experience shows this. Can drilling a well pollute drinking water aquifers? That's a whole different kettle of fish. It can pollute aquifers, but it doesn't have to.

Well integrity can be tested during and after construction, and remedial action can be taken to repair faults. In some jurisdictions, that is mandatory.

We have not spoken about well design here. Some wells are planned so that only the lower part of the casing is cemented into the rock, meaning the production horizons are isolated from the surface but most of the hole is left open behind the casing. Here is a quotation from one State's regulations governing the casing of coal seam gas wells "…*the production casing must be cemented in place with sufficient cement to allow for 200' of cement over the uppermost coalbed that the operator intends to complete."*

Let's think about that for a moment. State regulations accept an open hole section between the coal seam minus 200 ft (60 m) and whatever surface casing is cemented into place (the regulations are vague about that). Perhaps the life of the well will be 20 years (it could be much more or less) and throughout that life groundwater will be free to move vertically up or down the well bore. To make matters worse, this is a coal seam gas well and will probably be just one of many in this seam. Meaning the surrounding area will be peppered with similar potentially damaging wells. All with regulatory approval.

Can such wells pollute shallow aquifers? Absolutely, but that is a choice made by the voting community. There was a time when anyone rich enough to own a coal mine or a factory expected to pour waste water freely into the nearest river. You may know of rivers in coal mining areas that still run black with coal dust. A major river near my home used to run red-brown as a result of hydraulic mining for tin – and that was accepted as both legal and normal.

Times have changed and society no longer accepts willful pollution in the search for profits, but in some areas the polluters and their political friends have not yet noticed.

Frack Fluids introduce Toxic Chemicals into the Environment

This is a great concern to many people and needs careful consideration. Firstly, if we are talking about the fracturing fluid components, we have examined them in detail (see page 50 above). This list is largely benign but can contain small quantities of some hazardous materials – particularly as biocides or corrosion inhibitors. Much of the public worry about frack fluids comes from claims that they contain 500+ toxic components, but that is simply not true. You would not volunteer to drink a glass of a modern frack fluid, but you would probably survive the experience undamaged (please don't ask me to demonstrate that!)

Frack fluid should never be free to enter the environment and if you have been following the frack process described above, you have seen that the only way it can do so is by accident. It might be spilled on the well site, or a truck delivering it might leak. The same could be said for diesel fuel, and that is much more polluting. Frack fluid itself is not very dangerous and it has limited potential to be released into the environment, but it does have to be handled responsibly and plans made to handle possible spills.

Much more problematic is the fluid the well produces after it has been fracked. As part of the fracturing process, the spent frack fluid must be produced back to surface and what returns first is usually just that, frack fluid minus proppant and with the gelling agent broken. This is quickly followed by spent fluid with increasing reservoir fluid content, and the well operator has an immediate problem.

The difficulty is this; reservoir fluids include hydrocarbon liquids along with the formation water (normally brine rather than fresh water). What comes back to surface is basically a very large volume of contaminated salty water that has to be disposed of. The contaminants include the frack chemicals plus some potentially nasty components of the reservoir fluid – including the notorious BTEX suite of volatiles, benzene, toluene, ethylbenzene, and xylene. These are definitely not chemicals you want floating around in the environment. They are toxic and carcinogenic, and must be disposed of properly, even if they are only a tiny component of a large body of more or less salty water.

In short, frack fluids are unlikely to do environmental damage as a matter of course, although accidental spills are always possible. Such spills are not likely to cause major problems.

On the other hand, the flow back after the frack job is a potential cause of environmental contamination and, if we are not to rely on the operator's goodwill alone, we need strong regulations backed up with strong enforcement. The problem is predictable and can be dealt with.

What about the Halliburton Loophole?

The Halliburton Loophole is a way of referring to a series of exemptions from environmental regulations granted to the US oil industry by the Energy Policy Act of 2005. The Act is a wide ranging piece of legislation designed, among other things to introduce new safety standards, streamline licensing procedures and assist energy research. During the drafting and passing of the Act, Vice President Dick Cheney was active in inserting important exemptions for the oil industry from the Clean Air Act, Clean Water Act, Safe Drinking Water Act, National Environmental Policy Act, Resource Conservation and Recovery Act, Emergency Planning and Community Right-to-Know Act, and the Comprehensive Environmental Response, Compensation, and Liability Act (known as the Superfund Act).

Dick Cheney had been CEO of Halliburton, a well-known American oilfield service company involved in fracking and, when it emerged that hydraulic fracturing had been specifically included in the exemptions, this exclusion became known as the Halliburton Loophole. It has nothing to do with Halliburton as such and today the exemptions apply to all fracturing operations.

Of course, it does not apply outside the USA. Also, it is Federal Legislation and does not supersede the State regulatory requirements that govern day-to-day fracturing operations. In practice, it excludes the EPA and Federal Government from oversight of all aspects of fracturing. The power and duty to control them rests with the States who generally do not have anything like the expertise and resources to monitor potential damage. Most Americans have heard of the Environmental Protection Agency (EPA) but have little idea who covers that function at State level.

On paper, the Halliburton Loophole looks like a licence to pollute but in fact it simply passes control of fracturing to the State authorities who may or may not be able to resist the influence of the oil industry. How effective they are at their job ultimately depends on the State electors.

Note that the main battleground between well operators and State regulators is the disposal of water coming from the well. Fracked wells do not have to be dirty, but keeping them clean does cost money the operators might prefer to keep in their pockets.

Fracked Wells Pollute Rivers, Ponds and Lakes

The most troubling environmental issue for well operators is disposing of the water they produce along with their hydrocarbons. As we said above, in the bad old days of the mining industry, society considered it quite normal to run mine drainage straight into local rivers, no matter how much damage it did. This is still the case in some countries.

Modern oil and gas production is subject to much more stringent rules, including regulations governing the safe and environmentally responsible disposal of waste water. It can be a major part of the project cost, especially in the case of coal seam gas wells.

Produced water is typically too salty to be used for irrigation. Formation brines inevitably contain some heavy metals – including some toxic ones – in varying concentrations. They will probably also contain some hydrocarbon traces, including BTEX elements, which have not been completely removed by the separation process that prepares gas and crude for market.

Formation brines will probably contain naturally occurring radioactive minerals and these always cause concern in the general public. Brines will have a higher concentration of radionuclides than, say, drinking water simply because they have a far higher level of dissolved solids. Radionuclides can be brought to the surface as drill cuttings and sludge. They are also concentrated in scale built up on process pipework. Radioactivity is a natural phenomenon and surrounds us all the time but the radioactivity of formation brines might be two orders of magnitude higher than drinking water (but still far below hazardous levels). The oil and gas industry must always be aware of the issue, especially its effect on the disposal of waste water.

So – does fracking pollute surface water bodies? Absolutely, if we let it. Again, fracking does not have to be dirty. It can be as clean as we want and it is up to us to elect politicians who understand the problem and are prepared act for all society and not just the part of it that owns oil wells.

Fracking puts Methane Gas into Drinking Water

I think everyone has seen at least one video clip of a householder setting fire to the water coming from his kitchen tap. There can be no question that there are domestic water wells around that also produce gas. Flammable drinking water is a fact, not an urban myth. This has been known for many years but older records pre-date any fracking activity.

So – the gas flaming in the kitchen sink; did it come from a nearby fracked well, or is it simply naturally occurring shallow gas? Both types exist, and to confuse the issue, gas well drillers are attracted to locations with natural gas shows.

Can fracking introduce methane to a domestic water well? Probably that's the wrong question and we should be asking can drilling a gas well, fracked or unfracked, introduce methane to a domestic water well? Yes, definitely. It can happen and undoubtedly has happened.

Let's go back to the well casing laid along your town street (people must be getting really fed up with you by now…) If that casing has been properly cemented and sealed into the formation, there is simply no route for reservoir gas to reach drinking water aquifers. It just can't happen.

However, and here's where the problem usually lies, if only the casing across the reservoir has been cemented and there is simply casing hanging in an open hole for most of that long string of pipe, things are different. There is a highway towards the surface for any gas that escapes through a leak, or that by-passes a poor cement seal.

In a well-designed well that has been properly constructed and properly tested and monitored, there is no risk of methane in domestic water wells.

In a jurisdiction that accepts poorly designed, poorly constructed wells, there is every reason to expect gas to escape into surface aquifers. And don't forget, a string of casing in an open hole may be gas tight at the moment, but what will happen when the casing is degraded by corrosion?

Again, the responsibility for constructing a safe well lies with the operator, and the responsibility for approving that construction lies with the local regulatory authorities. Leaky wells don't just happen; they are allowed to happen. They are allowed to leak by the operators, by the regulators and by the ultimate bosses –We, The People...

Fracking destroys the Landscape
Drilling an oil/gas well and preparing hydrocarbons for market is an industrial process. Building an industrial plant has impacts on the surrounding area whether it is a gas well, a pig farm or even a shopping mall. So what are the likely impacts of a fracked gas well?

Wells that are part of the modern fracked shale expansion are likely to be drilled in a cluster, that is, several wells drilled from one location. To drill a well for shale gas, the drilling and fracturing operations need an area of about 5 acres (2 hectares). This area, known as a well pad, will have multiple wells drilled from it, perhaps 6 or 10, it varies with the reservoir properties.

The pad will need a holding pond during drilling and fracturing and this is usually an excavation lined with a waterproof membrane. This pond may be filled in once the wells are operating. The cluster will have a Christmas tree valve assembly on each well, and some form of manifolding to link the well outlets together. There is likely to be basic separation equipment to get most of the water out of the produced gas, and some on-site storage for that water – usually a closed tank – where the contaminated water waits for removal.

The pad will need an access road, a power supply, and an easement for pipe lines carrying produced fluids from the well. Once in production, the well head area will be fenced off, probably enclosing a big enough area to allow simple workover equipment to operate. The outer parts of the area will not be needed once the drilling rig moves off and these may be reclaimed to something close to their original state.

And that is it. The visual impact on the environment is comparable in size to a family farm complex, but much less visually comfortable.

One aspect of well sites that is often forgotten is that they are not permanent. Once the gas runs out – in perhaps 20 years or less – the wells will be abandoned. This is done by bringing in a small rig and filling them with cement. The well casing is then cut below ground level and the land returned to its previous state.

Do fracked wells destroy the landscape? They certainly have a significant impact on it, especially during initial construction. It is part of the cost society pays for having gas and oil ready at our fingertips. As the field matures the impact decreases and eventually the landscape will return to something like its original condition. How close to its original condition will depend on specifications imposed on the well operators by their lease and local regulations.

Fracking uses up Excessive Amounts of Fresh Water

If there is such a thing as a typical fracked well, in the US a ball park figure for the amount of water needed to get the well into production is 5 million US gallons (19,000 m^3). In 2011, it has been estimated that all the shale gas wells drilling in the US consumed a total of 135 billion gallons of water (511 million m^3). That certainly sounds like a great deal of water but, as always, it's a question of scale. In the same year, the mining industry consumed 1460 billion gallons or around 10 times as much. Livestock farming consumed about 5 times as much as the gas wells. Farm irrigation uses a massive 300 times as much. The shale gas industry consumed less than 0.1% of the water extracted for use in the US.

The big picture is that fracked wells consume a significant amount of water, but the amount is manageable in the context of what is already being consumed.

Of course, all water issues are local. Supplying water for a well in the arid parts of Texas is much more of a problem than it would be in, say, Pennsylvania. Local authorities in Texas need to be much more active in managing water supplies. That's their job and the shale gas industry just has to adapt to the conditions. They can do that by maximising the treatment and re-use of water, both flow-back water and produced water.

Because trucking in water or even laying pipelines is so expensive, the efficient use and re-use of water is in everyone's interest. In the local residents' interest because a scare resource is not being wasted, and the well operator's interest because he will not be wasting money buying and trucking in more expensive water.

Fracking Trucks will wreck our Roads

From drilling to completing a fracked shale gas well will typically need hundreds of truck movements bringing in equipment and supplies (including water) and removing waste. A well pad with, say, ten wells will certainly have thousands of truck movements over the construction phase – usually around a year. Most of those trucks are not unusually heavy or long, but there are a lot of them.

The difficulty is that no normal rural area has the roads, culverts and bridges to handle this sort of traffic, and when you are talking about hundreds of well pads being developed over many years, obviously something has to be done about this problem.

The development of a shale gas field imposes big infrastructure costs on host communities. The local authorities must understand the scale of the problem and plan for it. Gas producers need access for their trucks and, in the end, they must pay for it. Probably reluctantly, because who likes paying taxes? They are producing gas to make money and some of that money must be used to protect community assets like roads, bridges and culverts. The local authorities must make sure this actually happens because, for a large shale gas field, the trucks will be around for a while.

Fracking causes Earthquakes

Well, yes. Both directly and indirectly, that's a true statement but not a helpful one.

Anything that moves deeply buried rock around causes an earthquake of sorts. During fracturing operations, geophysicists can listen to the frack developing using microseismometers, but the 'quakes' that fracking generates are far too small to be noticed by anyone else.

Bigger quakes might happen if a gas reservoir is produced over a longer period. The rock is de-stressed as gas is produced and reservoir pressures drop. In time the rock might react by settling a little but, again, any such tremor is far too small to notice. This effect is not limited to fracked wells and it is likely that wells that produce more liquid (oil and water) will have more associated seismic activity.

Seismic activity around gas wells is miniscule compared to the activity around deep coal mines. In a deep mine, the rocks are destabilised by the removal of coal and large quantities of drainage water, and the ground regularly falls to fill the cavities left by the coal.

A bigger sources of earthquakes are water reservoirs on the ground surface. Creating an artificial lake behind a dam alters the pore pressure in rocks below and earthquakes are common, some of them strong enough to damage buildings and kill people.

The biggest sources of induced seismic activity are water injection wells. One way of disposing of waste water from oil and gas wells, and from other sources like mining, is to pump it back underground. Another reason to pump water underground is to increase reservoir pressures that are being depleted by hydrocarbon production, to encourage the remaining oil and gas to keep flowing.

There are tens of thousands of waste water injection wells working in the US and as they pump in water under pressure, they cause major changes in pore pressure underground. In geologically stable areas of the country, they may cause a continuous rumble of frequent micro-seismic events. In tectonically active areas like California the results can be much more dangerous by triggering natural seismic activity, possibly even devastating earthquakes. Adventurous scientists are experimenting with the idea of controlling earthquake size by encouraging many small quakes rather than allowing a single massive event to occur at random.

Oklahoma has seen a big growth in waste water injection over the last decade, and now has more than 100 quakes of magnitude 3 or greater per year. This size of quake is capable of causing damage and Oklahoma is learning to control and live with waste water injection. So fracking does not cause earthquakes, but disposing of the water fracked wells produce by injecting it underground might cause quakes.

Fracking brings Radioactive Gas into House Cellars

Radon gas is a decay product of the radioactive isotope radium. It is an active source of radiation in the environment, and dangerous to humans. The gas is found most commonly where there are suitable source rocks, mostly granites and glacial material derived from granite areas. It can collect in domestic basements and be a real long-term hazard.

Radon is working its way to surface everywhere so there is a natural background radon level that varies from place to place. If you live in a high radon area, like the Appalachians or Iowa, you should be aware of the problem and remove the gas by ventilating your basement. If you live in an area where radon levels are naturally low, like Oklahoma or Texas, you have much less of a problem. Check the EPA site at http://www.epa.gov/radon/find-information-about-local-radon-zones-and-radon-programs#radonmap for more details.

Can fracking increase radon levels in your basement? Possibly, but you would have to work out just how that could happen. One fortunate property of radon is its short half-life of 3.8 days. This means that natural gas entering the distribution system has a very low radon content by the time it is used domestically, and cooking or heating with gas makes no material difference to domestic radon levels. In some areas radon is present in the groundwater, so houses with their own wells may show measurably higher radon levels.

Natural gas as produced from gas wells contains a very small percentage of radon but if gas from a nearby well is leaking into your basement, you have other major problems to deal with before you worry about radon.

Radon gas is dangerous, but hydraulic fracturing has not been shown to increase that danger. The most significant source of domestic radon is the geology below the house and if you live in granite country you are more at risk than elsewhere. And natural gas is definitely not found in granite.

So – what could possibly go wrong?

Drilling and fracturing wells is a human endeavor so the honest answer to the question is – almost anything. If we refer back to the Deepwater Horizon blow-out, there was a catalogue of human ignorance and arrogance that is hard to believe. That well was very unusual, very deep, in deep water and challenging downhole conditions, so you would imagine that everyone was paying the closest attention to their work. How much more opportunity is there for error in developing a shale gas field with thousands of wells being drilled with a routine mindset?

Things can go wrong, just as airplanes go wrong. The way to stop or at least absolutely minimise problems is to pay close attention to details – just as in the aviation industry.

This is not an impossible task. The best well design and construction techniques are common knowledge but all too often doing things cheaply is more important than doing them right.

For a modern, safe industry, trained and certified staff are essential and they must work in a safe and regulated environment. Of course, there is a cost attached to that and this is where things get political. An industry cannot be expected to regulate itself. Some-one has to set the standards and then check continuously to make sure they are being met.

For far too long, dodgy behavior on the part of some operators has been accepted or even encouraged at the highest levels (see the Halliburton Loophole) and the challenge going forward is for the community to take control of what is sometimes a big mess.

There is one way oil and gas production differs from the aviation industry. Mistakes are rarely catastrophic. It is easy to check for casing leaks during construction, and repair them. It is easy to check the effectiveness of a cement job and do something about it. If a producing well develops a leak, modern telemetry means the operator should know about it within minutes.

If the worst comes to the worst, a properly constructed well can be shut in and abandoned safely.

There are no problems associated with the drilling and fracturing of wells that inevitably endanger people or the environment. It is a difficult process, but it is as safe and clean as we choose to make it.

Taking off the Gloves

I have been writing this book for a while now and really, really doing my best to talk engineering only, and to be neutral about fracking. Well, perhaps the last chapter got a little political, but it is still meant to be neutral. But of course, I do have opinions based on experience and study, so now is the time to put them down in writing.

I think the whole idea of being pro or anti fracking is just silly.

Hydraulic fracturing is simply a technique used in hydrocarbon production. There has recently been an oil find just south of London, England. The company concerned felt that in the coming battle with public opinion and the planners to develop the field, the idea of fracking would be toxic. They decided to voluntarily take hydraulic fracturing off the table and limit themselves to what they can produce by horizontal drilling only.

That's fine, but if you are a neighbor, do you really care if the well is fracked or not? The well drilling and construction will be pretty much the same. The visual and traffic impact over the next twenty years will be just the same. Neighbors will be objecting – presumably – because they would rather see a field with cows and horses than a well pad. They will be objecting to *any* hydrocarbon production because they do not want to make the compromises involved, and that is a perfectly valid position. Fracking has nothing at all to do with it, but the idea of fracking was a powerful political argument that had to be dealt with.

Oil and gas production is a mining process, producing minerals from the ground. As mining goes, its impact is relatively light – a coal mine does much more damage. But oil and gas wells do have an impact. What should concern the average citizen is whether or how he is going to live with it.

Learn, Learn, Learn...

There is a very vocal lobby against fracking, and a very rich and powerful one supporting it. How are you going to get your opinion onto the table when decisions are being made?

The main essential is that you know and understand what you are talking about. Here in Australia we have an influential anti-fracking lobby group called Lock the Gate. Their car stickers are everywhere and if they call a local meeting, you can guarantee it will be well attended and have good press coverage. But I bet not one in a thousand of its supporters know as much about the topic as you do after struggling through all the engineering in this slim book. They are an organisation of well-intentioned people who are strongly against fracking but really don't understand why.

Does that matter? You can bet it does, because they are not going to win every battle. The oil companies are just too big, and they *do* know what they are talking about. Anti-fracking activists need to be able to ask the right questions, and understand the answers they get. They don't need to waste resources highlighting the dangers of radon gas causing birth defects in mice because, while they are losing that battle, the potential producer may be slipping through cheap and nasty casing designs that nobody understands.

Both supporters and detractors need to properly understand what they are talking about because poor arguments lose battles.

Where to Learn?

In the 21st century, our first stop will always be the internet but straight away we run into the classic internet problem. There is just so much *stuff* out there, and most of it is there to persuade, not educate.

That's not so bad, as long as you can sort out the agenda behind what you are reading. The easiest ones to spot are the oil industry explanatory pieces. Go onto Youtube and you will find some beautiful animations of drilling and fracturing, and most of them are pretty good, as far as they go. Of course, the oil companies have paid for them because they want you to feel comfortable with the fracking process and confident that it is in the best and safest of hands i.e. theirs.

In the opposing camp, there are many less well funded pieces from the anti-fracking lobby, and here life gets really difficult. Some of the pieces appear to be made by obsessive crackpots who start from a position of 'all fracking is bad' and will use any method they can to get you to join their camp. Others are more measured but are still pushing an anti-fracking agenda rather than stepping back and looking at the whole topic. Mixed in with the hype, there are sad stories of people who have had their groundwater destroyed by badly constructed wells but these videos concentrate on the damage done to people rather than investigating how exactly it happened.. Or what regulatory failures allowed it to happen. Not much can be learned without some sort of technical investigation, and future failures cannot be prevented unless we know exactly what caused the current problems.

And then we have the media, and some of them are terrible. We are still getting presenters on radio and television explaining how fractures are made by lowering explosives into the well and blasting cracks in the rock. Even the more responsible outlets fall far short – here is CNN explaining what fracturing is all about

https://www.youtube.com/watch?v=LAxsTJd7VCA

Having read this book and looked at the CNN clip, I think you will agree that no-one is going to learn enough about fracking simply from Youtube.

Where else? Should we be reading about fracking? Of course, I am going to say yes. I read and write, I like books as a way of conveying more detailed information. Unfortunately, searching for basic information on fracking on Amazon does not yield much. There are some self-published opinion pieces, none of which I can recommend (there may be good books I did not find, of course.) Again, they are mostly about the effect on individuals of wells that have gone wrong. You now know that simply blaming fracking for contaminated water wells does not help, but people with no drinking water need to blame some cause. The people who may have caused the damage – the well owners – will go into hiding. They certainly will not help writing a book about how the problem happened, and without their input, a farmer's story of his ruined water well will be heart-breaking but we will not learn anything useful. The general reader wanting to know more about fracking is not well served.

The Films

There is a landmark film that has had immense success and is responsible for shaping the policies of many governments when it comes to hydraulic fracturing. It went on general release – unusual for a documentary – and even financed a sequel. Unfortunately, it also qualifies for the Dr Goebbels Golden Globe Award for Propaganda.

The film I am talking about is, I am sure you have guessed, *Gasland*. If you have not seen it, you must watch it and form your own opinion. For me, it is intended to manipulate its audience from the beginning and right the way through. It is simply a political instrument aimed at manipulating public opinion. And a very successful one because it provoked a wave of anti-fracking sentiment in the general public and was at least partially responsible for blanket bans on hydraulic fracturing in some states and even countries.

An example of its unashamed use of propaganda techniques is the sequence showing a man lighting the water flowing into his kitchen sink. This is probably just what it seems to be – gas in a domestic water supply – but the film is not at all interested in why the gas is present. Presumably it is either naturally occurring shallow gas or a substantial leak from a badly constructed well, but the film is not concerned with isolating the source. It simply uses the image to re-inforce its anti-fracking message – allow fracking anywhere near your house, and you too will be able to light your drinking water.

Do watch *Gasland*, but immediately afterwards you should watch another film, *Fracknation*. This is also a documentary film with an agenda, and that agenda is to ask the difficult questions about *Gasland* and it does that very well. Neither film will help you learn much about fracking.

A film that is definitely worth watching is *The Facts on Fracking*, a filmed lecture by Dr Anthony Ingraffea. Dr Ingraffea is a very knowledgeable man indeed and was right at the forefront of the theoretical physics of fracture development. His lecture is 1-3/4 hours long and should be required viewing for anyone aspiring to hold an opinion on fracking. Interestingly, although he is a leading expert on fracking, he does have reservations over what is happening in the US today. To summarise his conclusions, he is concerned about the quality of regulatory control, and about the scale of current operations (how often have you read about scale just recently!). Anyway, do watch the film and judge for yourself.

So where does that leave us?

Everyone's personality is made up of a compost of biases, prejudices and opinions, and nothing they say or write should be taken entirely at face value. That includes me, of course.

At least, I no longer have a personal stake in the oil industry, and no-one is about to drill wells anywhere near me. If I have an agenda, I suppose it is to educate people about fracking. I find the level of ignorance in the media personally offensive. I don't shout at the television or throw the remote at it, but I have been tempted!

Anti-fracking obsessives also annoy me with some of the barely concealed rubbish they put on the internet, but what can we do about them?

I decided the best I can do is to write something that is factual and readable to help people form their own opinions. I know I have managed the factual part; but the readable bit... It is difficult to make the details of well design interesting to the general reader but if you have got this far, I am sure you will agree that you cannot understand fracking without a basic understanding of how wells are constructed.

Well done, you. I'm sure you deserve some sort of certificate. In fact, getting this far definitely entitles you to a certificate and if you email me, I will send you one. It will be a pleasure to meet some-one who cares enough about fracking to study the details.

About the Author

Eric George studied as a geologist and then moved into the oil industry as an engineer. He worked in many countries and became and expert in high pressure pumping and cement technology. He has now retired from the rigs and lives a comfortable life in north Queensland, on the shores of the Coral Sea.

About the Writer

Jacqueline George lives in Cooktown, a village in the far north of Queensland, Australia. When she is not cutting the grass or taking cake and coffee in our little patisserie by the Endeavour River mouth, she passes her time writing. Some of her books are fun but hardly respectable. Others are more serious (but still fun). You can find all her titles at

www.jacquelinegeorgewriter.com

CPSIA information can be obtained
at www.ICGtesting.com
Printed in the USA
LVHW02s2110201217
560369LV00022B/3125/P